LICENSE TO ORBIT

The Future of Commercial Space Travel

Joseph N. Pelton

and

Peter Marshall

LICENSE TO ORBIT

The Future of Commercial Space Travel

Joseph N. Pelton

and

Peter Marshall

ACKNOWLEDGEMENTS

The authors wish to express their special appreciation to Brett Alexander, Peter Diamandis, John Gedmark, Rob Godwin, Michael Gold, Mark Goodman, Nick Goodman, Alexander and Eloise Pelton, Tomasso Sgobba and Mark Williamson for their assistance in so many ways in the preparation and production of this book. As always, any errors are those of the authors.

Published by Apogee Books, and imprint of Collector's Guide Publishing Inc.,
Box 62034, Burlington, Ontario, Canada, L7R 4K2
http://www.cgpublishing.com
http://www.apogeebooks.com
Printed and bound in Canada

License to Orbit – The Future of Commercial Space Travel
ISBN: 978-1-894959-98-8 – ISSN: 1496-6921
©2009 Apogee Books

TABLE OF CONTENTS

INTRODUCTION

By Dr. Jeffrey Hoffman

(Photo courtesy of NASA)

Forty-five years after the first airplane flight, as humanity emerged from World War II, air transportation was becoming a widespread commercial activity, with a multitude of privately owned and government-supported airlines on all inhabited continents. The advent of commercial jet travel, which has revolutionized social and commercial activities throughout the world, was just on the horizon.

Forty-five years after the launch of the first artificial satellite, human civilization has been transformed by space telecommunications, navigation, weather forecasting, global monitoring, and scientific research.

Forty-five years after the first human space flight, however, only three government-operated space agencies have the capability of sending human beings into orbit, and the impact of human space flight on global civilization has been mostly indirect, in the inspiration it engenders and the technological development it fosters.

Can human-carrying spaceships someday affect global civilization as significantly as airplanes and satellites? We will never know until it actually becomes possible for large numbers of people to leave the Earth and travel into space. When will that be?

During its most productive years, NASA Space Shuttles were carrying around fifty people a year into low Earth orbit, where they carried out a wide variety of activities such as science and engineering research, repairing satellites, and constructing the International Space Station. But fifty people per year is an insignificant number compared to the volume of people flying on even one jumbo jet!

Until much larger numbers of people have access to space, we will not know the true potential of human space flight. That is where the dream of commercial space travel, the "new space movement," gets its inspiration – to turn space flight into an activity that is accessible to large numbers of people.

The X-Prize for the first commercial sub-orbital space flight, won in the fall of 2004 by SpaceShipOne, attracted for the first time large-scale public attention to the idea of commercial space flight. It is now possible to pay a few thousand dollars to Zero-G Corporation in order to experience over five minutes of weightlessness, albeit broken up into twenty-five second chunks. Virgin Galactic's SpaceShipTwo and other commercial space initiatives promises to give this amount of weightlessness all in one piece and throws in a breathtaking view of the Earth seen from the blackness of true space, all for about a quarter million dollars.

It will take almost fifty times more energy to get into orbit, however, and currently only the Russian Space Agency is willing to provide this experience commercially, to a very limited and extremely wealthy clientele. However, we are living in a time where quite a few billionaires who were inspired by Apollo are devoting their own resources to making the dream of "space travel for the masses" a reality. It is not going to be easy, as several entrepreneurs have already learned. But these people have both engineering and financial acumen, and although building rockets is, indeed, "rocket science," at least some of them will hopefully succeed in being able to launch people into orbit and take care of them in private space stations for a reasonable enough price to create a whole new industry.

How do I, as a NASA astronaut who flew five Space Shuttle missions, feel about this possibility? Well, it will certainly change the nature of what it means to be an astronaut! But I truly welcome the possibility that private enterprise might be able to support and maintain the infrastructure to launch people into space and take care of them once they are in orbit, because I believe in space exploration and I see private space flight as a necessary step in the eventual expansion of humanity into space. Right now, NASA spends over a third of its entire budget maintaining this infrastructure, and if it could instead buy these launch and orbital housing services from the private sector at the marginal cost instead of having to maintain the entire infrastructure, it would have far more resources to devote to true exploration, which is what I think the goal of government-supported human space flight should be.

The path to "new space" is only just beginning and, whether you are a believer or a skeptic, License to Orbit by Dr. Pelton and his colleague Peter Marshall is a great read. It provides clear insight and perspective on the technological, legal, economic and social challenges that face the new generation of commercial space pioneers.

Jeffrey Hoffman is a scientist who joined the NASA space program as an Astronaut in 1978. He went on to make five Shuttle flights between 1985 and 1996, including the repair of the Hubble Space Telescope. He logged a total of 1,211 hours and 21.8-million miles in space. He is now a Professor at the Massachusetts Institute of Technology (MIT).

— Chapter 1 —

"I believe that the Golden Age of space travel is still ahead of us. Before the current decade is out, fee-paying passengers will be experiencing sub-orbital flights aboard privately funded passenger vehicles, built by a new generation of engineer-entrepreneurs with an unstoppable passion for space We are seeing the emergence of a new breed of 'Citizen Astronauts' and private space enterprise." — Sir Arthur Clarke.

CITIZEN ASTRONAUTS: HOW SOON?

Are you ready to buy a ticket and go flying off into space? How about spending a few days aboard the International Space Station and viewing the Big Blue Marble we call Earth from several hundred miles up in space? No way, Jose! At least that is what most people you meet on the street would say if you were to ask them right now.

Since the cost of such a junket comes out to be in the tens of millions of dollars, that's an easy one for most people. But the price of so-called space tourism is coming down – sharply. Virgin Galactic and other private companies who are now developing space planes for sub-orbital flights promise that they will shortly offer a brief flight into outer space for much less. Most space tourism companies indicate their prices for 2010 or so will be somewhere in the range of $200,000 to $250,000.

That's still too steep – perhaps both financially and physically – for most of us. We peons that fly tourist class and think that tickets to fly should be calculated in hundreds of dollars rather than hundreds of thousands are not the prime target market here. Today, the space tourism companies are busy trying to sign up millionaires, celebrities, royals, rock stars, movie stars and star athletes with money to burn on the rocket fuel *du jour*. The key to the business plans of most private space entrepreneurs is to get the passenger volume and the space planes UP and the prices DOWN. The truth is that times do change. Travel by airplanes used to be expensive, risky and just for the elite. The space tourism business hopes that they can repeat what happened in the aviation industry.

In the first half of the 20th century, what would any sensible person have said to the idea of booking a flight from New York to London for a weekend jaunt? Or how many people would have said: "Oh yes, one day we will be buying hand-held electronic computers for a few hundred bucks? And by the way, just tell me exactly what is an electronic computer and exactly how do ordinary people use it?"

Back in the 1940s you might have asked people about their views on lasers, transistors, satellites, bio-engineering, and cloning of animals. Their reaction would have been much the same. They would have simply stared at you and wondered what asylum had let you loose on an unsuspecting world.

Manned rockets to the Moon? The Internet? Even spandex? All these things and more would have been seen as sheer science fiction just a few decades ago. But times change. Science fiction becomes harder and harder to write simply because technology keeps accelerating us to more and more improbable levels. Today, that change comes at us all too, too, fast! It's a crazy world where genetic engineering exists in a world right along with near Stone Age people in remote villages living on subsistence farming or nomads living off the land.

We used to say: "The sky's the limit." But that is no longer true. Now we can all sign up to become Citizen Astronauts (an apt name coined in 2007 by the world renowned scientist and science fiction writer the late Arthur C. Clarke). But there are a few catches. First, we must be willing to wait a bit for our ride to space. Also we have to sign some heavy-duty waivers. These will say in legalese that we know that a sub-orbital flight – going up some 60 miles or so into outer space on a commercial space plane, and then down – is a very risky and still a very experimental flight. We know we may be killed and that our estate will have no grounds for collecting damages because we have waived all rights to liability claims against the company or the government.

Thirdly, after waiving all rights to seek damages and shelling out $200- 250,000, we must then go through training for a very rigorous "high g" ride that lasts only a few hours to experience three to four minutes of weightlessness and only a very brief time in outer space. Today's space tourism is clearly not for everyone.

But within a decade or so, as the new space tourism business develops, travel through the stratosphere might not even break your piggy bank. Today, rocket engineers are seriously working on new technologies that will enable more and more people to become Citizen Astronauts. Their mission statement is quite simple: to make space travel a lot cheaper and not nearly as scary.

Such a mission statement is key to the success of the new personal space flight companies. No real market will develop until space flight becomes not only affordable, but also a lot safer than it has ever been before. It may never be as safe as riding a passenger jet, but that is the objective of the industry leaders behind the newly-formed Personal Spaceflight Federation (of which more later).

Despite the high prices and major risks, this new space travel industry will soon be the new fashion. In just a few years, it is expected that thousands of fare-paying passengers will fly more than sixty miles (or 100 kilometers) into space on board the next generation space planes. If these entrepreneurial companies establish a new commercial space business with some success, a trip to outer space will take on new meaning. These citizen space travelers could, within a decade, far outnumber the 500 or so astronauts and cosmonauts who have gone into space so far. In another two decades literally tens of thousands of people – not just trained astronauts or wealthy celebrities – might be able to experience weightlessness and see the world from the perspective of space. Ordinary people will start to follow in the footsteps of cosmonaut Yuri Gagarin and astronauts John Glenn and Neil Armstrong and the other history-making spacefarers who led the way in manned spaceflight. This will be the new era of Citizen Astronauts as we enter a new Commercial Space Age. In short, Fantasy could become Reality.

But who is making this all possible? Is it NASA or the European Space Agency or the space agency of some other country like China or Japan? No! Rather amazingly it is a handful of 'Space Billionaires' who have already made their fortunes in other areas as such as computer games, music, hotels and the Internet. These entrepreneurs with "GIGABUCKS" are now making all this possible; and what they have in common is entrepreneurial talent and gutsy, bet-the-company risk-taking.

What journalists called "space tourism" truly began on April 28, 2001. This was when the American industrialist Dennis Tito spent over $20-million to board a Russian Soyuz spacecraft for a rocket ride into space to visit the International Space Station. Since then, a total of six Citizen Astronauts have slipped the bounds of Earth's gravity by mid-2009. The fourth of these was the first female private space tourist, Anousheh Ansari.

Suddenly, private sector initiative is enabling "mass access to space." These private commercial ventures hope to make civilian access to space occur much faster than a NASA or other officially constrained governmental space agency could ever contemplate. We are now on the threshold of a new era of space tourism almost undreamed of a mere decade ago.

Figure 1.1: Professor Stephen Hawking – Freedom from his wheelchair to experience Zero-G.
(Photo courtesy of Zero Gravity Corporation)

Yet another milestone in this quest to reach the high frontier was a remarkable "weightless flight" by the famous British physicist and cosmologist Professor Stephen Hawking. The professor overcame his severe disabilities with motor-neuron disease (known in the US as Lou Gehrig disease) to experience weightlessness on a Zero-G flight. Afterwards, he said: "Space – here I come." Apparently he found it exhilarating to escape for just a short while from his wheel chair. Hawking thus declared his ambition to make a sub-orbital flight with Sir Richard Branson's new space travel company, Virgin Galactic, which plans to be in service in 2010. Sir Richard has agreed to "fix it" for the professor and to arrange a Virgin Galactic flight for him.

In more prophetic mood, Hawking also told "The Statesman," a British publication:

> "I am hopeful that if we can engage this mass market, the cost of space flight will drop and we will be able to gain access to the resources of space and also spread humanity beyond just an Earth-based existence. Sooner or later, some disaster may wipe out "intelligent life" on Earth."[1]

Algae and lichen might survive but nuclear or biological weapons or a meteorite or comet could conceivably extinguish humanity. In the view of Hawking and many other noted scientists, the long-term survival of the human race requires that we spread our "seeds" into space.

The First Five

Dennis Tito, the first fare-paying astronaut, started his career in the industry as an aerospace engineer in NASA's Jet Propulsion Laboratory in 1963. There, he calculated trajectories for interplanetary probes of the Mariner series. Tito is therefore not only a long-term space junkie, but also a very well-informed one. As a space scientist he participated in the robotic exploration missions to Mars and Venus. Then, in 1972, he founded his own company Wilshire Associates Inc. in Santa Monica, California, which went on to become a leading provider of services in the areas of management consulting and investment technologies.

Dennis Tito's enthusiasm for space travel continued unabated over the decades, and led him to pay something over $20-million for the privilege of becoming the first paying customer in space. His epic trip was arranged by Eric Anderson, another space enthusiast, who gave

Figure 1.2: Dennis Tito returns from his historic flight
(Photo courtesy of Space Adventures Ltd.)

up believing that NASA might be able to think outside the box and started his strive to bring space to the masses. He decided that access to space on a broad level could best be attained through private enterprise and left the security of a NASA job to become a high-flying entrepreneur by founding Space Adventures Ltd.

He boldly went where no one had gone before. In spite of NASA's rather predictable bureaucratic objections, he managed to broker a deal with his newly-found Russian partners to begin ferrying space tourists to orbit. The cash-strapped Russian Federation's space agency agreed to sell a seat on their Soyuz vehicle that was slated to go to the International Space Station. But this was not like booking an airline ticket to Moscow.

Tito actually began training to be the first citizen astronaut at the Gagarin Cosmonaut Training Center in October 2000 for his flight to the International Space Station (ISS). After rigorous training at the Star City facilities, just outside Moscow, he was judged fully trained and competent to make the trip to low earth orbit. On April 20, 2001 he blasted off for his journey into the annals of history. Despite NASA's objection that only fully trained astronauts and cosmonauts could fly into orbit, he discharged his duties competently and returned safely from the 8-day mission. After his return, the American businessman appeared before a US Congressional Committee to tell his story, in which he recalled:

> "There was one thing not even the most extensive training could prepare me for: the awe and wonder I felt at seeing our beautiful Earth, the fragile atmosphere at its horizon and the vast blackness of space against which it was set. Just imagine being able to watch 16 sunrises and sunsets each day. And, thanks to a team of generous ham radio operators and the crew on the ISS, I was able to connect more clearly with my sons down on Earth than I had previously when we were face-to-face. As any one of the 400-plus people who have traveled to space will tell

you, no amount of training can prepare one for the experience of weightlessness and the freedom of effortless movement. It remains something that's still hard to describe to others. I can say that you get a sense of total relaxation. The nights I slept in space were the best nights' sleep I've had since I was a baby."[2]

Dennis Tito's space flight was followed in April, 2002 by a South African computer millionaire Mark Shuttleworth. Although born in Welkom, Free State, South Africa, he currently lives in London and holds dual citizenship of South Africa and the United Kingdom. He spent more than a week on the International Space Station participating in experiments related to AIDS and genome research. To participate on the flight, Shuttleworth had to undergo one year of training and preparation, including seven months at Star City. Just as Tito before him, he recognized he was not only purchasing the chance to do undergo a unique experience afforded few humans, but also buying a place in history.

Figure 1.3: Mark Shuttleworth boarding the Soyuz spacecraft (Photo courtesy of NASA)

Figure 1.4: Gregory Olsen (center) and his Russian colleagues in orbit
(Photo courtesy of NASA)

The third fare-paying space traveler was Gregory Olsen in October 2005, who like Tito was a trained scientist and whose company produces specialized high-sensitivity cameras. Olsen cleverly conceived his trip to the International Space Station in the context of promoting his cameras and his company. Olsen thus used his time on the ISS to conduct a number of experiments, and also in part to test his company's products.

By this time, the story of fare-paying astronauts had started to appeal to the journalists who report on outer space as an ongoing story. But to some writers, these aspiring "spacemen" were beginning to seem like no more than a bunch of extremely rich old guys willing to pay a fortune to realize a childhood dream. In the early days of space exploration, the special quality required for an astronaut was defined as "The Right Stuff" – the title of Tom Wolfe's 1979 book and the movie made in 1983. Now, it seemed that "the right stuff" was a large bagfull of dollars!

But in 2006, Citizen Astronaut number four definitely created a new series of headlines. This time the wealthy high tech guy in space was a very attractive, high tech young woman who did not even look her 39 years of age.

Anousheh's Dream

It was Anousheh Ansari, chairman and co-founder of the Texas-based Prodea Systems, Inc., who realized her dream of an ascent to Earth orbit. She is a member of the Ansari family that funded the $10-million Ansari X-Prize competition for manned space flight in 2004. This challenge did as much as any other single factor to fuel the furious rush to create viable, safe and much lower cost access to space.

Anousheh, who not so long ago graduated with flying colors from George Washington University, has done a lot in her short life. She and her husband founded a revolutionary digital home

Figure 1.5: Anousheh Ansari calls home from the International Space Station
(Photo courtesy of NASA)

technology company which, not too surprisingly, is now involved in developing a line of sub-orbital space vehicles. In partnership with Eric Anderson of Space Adventures, her company is now also backing two spaceports – one in the United Arab Emirates and another in Singapore. Another partner in these ventures is Peter Diamandis, the founder of the X-Prize, co-founder of the International Space University and now head of the ZeroG Corporation. Ansari, Anderson and Diamandis are the "anything is possible" type of entrepreneurs, hell-bent on making space tourism a reality and sooner rather than later. The three of them have all become a new kind of space rock star.

It was in September 2006 that Anousheh flew aboard a Soyuz TMA spacecraft from the Baikonur Cosmodrome in Kazhakstan, en route to the International Space Station. On her 10-day mission NASA astronaut Michael Lopez-Alegria and Russian cosmonaut Mikhail Tyurin were her companions to conduct a series of experiments on behalf of the European Space Agency (ESA).

After her return, Anousheh told *Space Futures*:

> "Fly[ing] to space and see[ing] Earth as a planet instead of a city or a country changes the way you look at things … Instead you start looking at everything differently and from a bigger perspective. You realize that you're part of a bigger universe, but at the same time you realize how vulnerable and fragile you are … You just think how silly it is that people on Earth fight for small pieces of land or things that seem unimportant."[3]

She added:

> "I would like to thank Space Adventures for providing the flight opportunity, the crews of Expedition 13 and 14 who made me feel very welcomed during my time spent aboard the ISS and all those in who helped me prepare for this adventure. I hope those around the world who followed my mission consider what their own dreams are and pursue them, as I have done with mine."[4]

Numbers five and six

Then, in April 2007, Space Adventures Ltd., made arrangements for Charles Simonyi, Ph.D., to be the fifth paying passenger to fly onboard a Soyuz mission to the International Space Station. The total price for his 13-day mission to visit the orbiting ISS was not revealed but it was reliably reported by Space.com to be in the $25-million range. As a software pioneer who developed some of the early Microsoft systems, Dr. Simonyi has laid claim to being the "First Nerd in Space." This mission grabbed a quite a few headlines, but the press focused mostly on the romantic angle. This was because Citizen Astronaut Nerd Simonyi is also the boyfriend of the highly successful American media chef and home designer Martha Stewart. It was reported that Martha was busy preparing duck pates and other exotic foods for Charles' return to earth.

Figure 1.6: Charles Simonyi – the 'First Nerd' in space.
(Photo courtesy of NASA)

However, before Simonyi's voyage as a Citizen Astronaut, there was also a serious element in his ascent to orbit. He was actually trained to assist several international space agencies by conducting experiments during his nearly two weeks in space. Like his predecessors, he carried out his final preparations for the mission at the Yuri Gagarin Cosmonaut Training Center in Star City, Russia, and then returned again for a second mission in March 2009.

Number six in the list of fare-paying passengers was Richard Garriott, a 46-years-old British-born American citizen. He is another successful businessman, who made his personal fortune working in the design and development of computer games. His 10-day flight to the International Space Station in October 2008, was inspired by his ambition to follow his father, Owen Garriott, who made two space flights as a NASA astronaut in 1973 and 1982.

Unlike Space Adventures' earlier clients, Garriot also participated in experiments conducted by the astronauts on board the ISS. "I am enthusiastic to participate in these experiments" he said before his flight. "As my father was a NASA astronaut, it seems fitting that I, as a private astronaut, also assist in research as a continuation of my family's contribution to the space agency."[5]

Garriot agreed to be tested for three experiments while he is in space – how eyes react to low and high pressure in space, the effects of spaceflight on the human immune system, and the sleep and wake patterns and characteristics of astronauts. He also brought back to earth fresh blood samples from the long-duration members of the ISS crew.

The first Citizen Astronauts were reported to have paid $20- to $25-million but the price has apparently now risen to $30-million due to the weakness of the US dollar against the Russian ruble. Space Adventures Ltd., the driving force behind this new business, reports that despite the cost, there is still a list of adventurers just waiting for the opportunity to experience the flight. The company is headquartered in Vienna, Virginia with "satellite" offices in Cape Canaveral, Florida, Moscow, Russia and Tokyo, Japan. It offers a range of programs with the high-end option being the spaceflight mission to the International Space Station.

But they are not stopping there. In mid-2007, Space Adventures announced that negotiations are underway for the first private human flight to orbit the Moon. According to their web site, two space tourists, each paying approximately $100-million, would ride with a Russian cosmonaut aboard a modified Soyuz spacecraft. The Soyuz would get a boost from a docked Russian upper-stage rocket and fly a "boomerang" trajectory around the Moon.

Figure 1.7: Richard Garriott in training at Star City, Russia
(Photo courtesy of Space Adventures Ltd.)

Then in July 2008, the company's founder and CEO Eric Anderson announced the first charter of an entire Soyuz flight. This first-of-a-kind space tourism trip will include only one professional cosmonaut, the pilot, who will ferry two spaceflight participants, or space tourists, up to the International Space Station. The announcement was made at the Explorer's Club in New York City on the ten-year anniversary of the company's founding. Anderson told his audience that this first private charter space flight " … showed people around the world that there was a market."[6] However, the recent cooling of relations between the US and Russia over the sometimes brutal military intervention in Georgia raises questions about not only the private uses of the Soyuz flights, but the also US reliance on Soyuz to get American astronauts to the ISS as well.

Space Adventures is a company that is always pushing the envelope, literally and figuratively. They are all the time developing new programs, including working with leading universities to carry out new space experiments. A colleague recently remarked that the space tourism business will be highly dependent on what comes next. One key certainly does seem to be having answers for potential customers who ask the "What next?" or the "And then what can I do?" questions.

There is a wealth of experience on the Space Adventures advisory board including the Apollo 11 moonwalker Buzz Aldrin, Shuttle astronauts Sam Durrance, Robert Gibson, Tom Jones, Byron Lichtenberg, Norm Thagard, Kathy Thornton, Pierre Thuot, Charles Walker, Skylab astronaut Owen Garriott and the Russian cosmonaut Yuri Usachev. Eric Anderson commented:

> "Ten years ago many people thought that space tourism was in the realm of science fiction. However, in 2001, through innovative thinking and pushing the boundary of what is possible, Space Adventures sent the first private citizen to the International Space Station, proving that commercial spaceflight is a reality. In the next 10 years, Space Adventures will fly more people to space than have flown in all of human history. We will create astronauts by the thousands, not by the one or two."[7]

Space Adventures announced in 2007 that another of their future passengers will be Adnan Al Maimani, to become the first "national" from the United Arab Emirates (UAE) to go into space (Anousheh Ansari was born in Iran and her husband is a citizen of Dubai). Further, it is planned that he will launch from the future commercial spaceport to be located at the Ras Al-Khaimah International Airport. The company had earlier announced plans to locate a spaceport in the U.A.E. that will be funded by various parties, along with shared investments by Space Adventures and the government of Ras Al-Khaimah. Also, Sheikh Saud Bin Saqr Al Qasimi, along with the U.A.E. Department of Civilian Aviation, has granted clearance to operate sub-orbital space flights in their air space.

But was Toyohiro the First?
But who really was the first paying passenger to travel into space? Ten years before Dennis Tito's ground-breaking space flight, the Russian space agency decided to allow Toyohiro Akiyama, a reporter for the Japanese television company Tokyo Broadcasting System (TBS), to fly to the Mir space station and return a week later, for a price of $28-m. Akiyama gave a daily TV-broadcast from orbit and also performed scientific experiments for Russian and Japanese companies. However the cost of the flight was paid by his company, which made Akiyama a sort of "business traveler" rather than a space tourist.

There were also earlier female astronauts many years before Anousheh Ansari, including the US schoolteacher, Christa Corrigan McAuliffe on the ill-fated Challenger mission in 1986. Then, in 1991, a British chemist Helen Sharman was selected from a pool of public applicants to be the first Briton in space. As the U.K. had no manned spaceflight program, the arrangement was made by a consortium of private companies which contracted with the Russian space program. Sharman was in a sense a private space traveler, but she was also a working cosmonaut with a full training regimen.

The Next Money-Maker in Space?
But it is the paying passengers like Tito, Shuttleworth, Olsen, Ansari, Simonyi and their successors who will serve as the catalyst for developing the sub-orbital personal spaceflight market that is likely to be the next money-maker in space. These were the wealthy enthusiasts willing to pay the high cost of an 'ego trip' with the proven Russian space program. But now, as the new breed of space entrepreneurs makes plans for their business enterprises, the cost for the next group of space travelers is coming to a level

COMPUTER RENDERING COURTESY ELRO.COM

Figure 1.8: New Space Ports Are Envisioned As Exotic in Architecture
(Artist's impression, courtesy of Space Adventures Ltd.)

closer to that of other exotic vacations. Several of the start-up companies are now working on vehicles that could give passengers about 4-minutes of microgravity and an edge-of-space view of Earth for about $200,000. They include Sir Richard Branson's Virgin Galactic, which is already building a spaceport in New Mexico for its commercial SpaceShipTwo – a follow-on version of the 2005 X Prize winner, SpaceShipOne, developed by Burt Rutan and Paul Allen.

A couple of purse-string minded misers will probably say: "Hey, $200K is still a big dent in the budget, do you have anything in the economy section or the cargo hold?" Space Adventures has developed quite a few options for those space enthusiasts who don't feel comfortable laying out a few hundred thousand for a vacation. These options start with a trip from either Florida or Las Vegas aboard the so-called "vomit comet" as provided by Peter Diamandis' ZeroG corporation – and as experienced by Professor Stephen Hawking. This provides just under a minute of weightlessness via a high parabolic flight, for just $3495. For rather more dollars, you can train to be an astronaut at Star city, simulate an entire trip to space, take a flight on a Soviet Foxbat at supersonic speeds to the true edge of space, or consider other options including that ride to the space station. That is if you can get a booking.

It is not only Space Adventures that has a waiting list. Applications to become one of the first 100 members of Virgin Atlantic's "Founder's Club" for flights on SpaceShipTwo, is now closed out.[8]

Meanwhile, speculation as to who will be the next celebrity to go into space remains rife with possibilities. The names of Hollywood actresses Victoria Principal and Sigourney Weaver have been reported, together with 'A British Royal' (reported to be Princess Beatrice, 20-years-old grand-daughter of Queen Elizabeth II, whose boyfriend works for Sir Richard Branson's Virgin Atlantic company). And Charles Simonyi, who has worked closely with Bill Gates since the start of Microsoft, has suggested that Gates may indeed be among those next to spring for a $25-million ride to space within the not too distant future. So far, Space Adventures will only say Gates is not currently booked.

This new market to make the ticket price of space tourism at least two orders of magnitude less (i.e. down from $25-million to $200,000 or so) was clearly sparked by the $10-million Ansari X Prize that Burt Rutan's Scaled Composites won with its air-launched, hybrid-fueled SpaceShipOne in 2005. But the SpaceShip Corporation founded by Rutan, his financial backer Paul Allen of Microsoft fame, and Branson's Virgin Atlantic have competition. The others in the running, described more fully in Chapter 3, include Blue Origin, a start-up backed by Amazon.com founder Jeff Bezos. His company plans to launch a vertical take-off and landing vehicle from a spaceport in West Texas. Another key player is Elon Musk's Space X Corporation. Musk, who made his billions by founding the PayPal Corporation, was awarded one of the two contracts in NASA's $500-million Commercial Orbital Transportation Services (COTS) program to develop a commercial re-supply vehicle for future astronaut transport to the International Space Station. The idea is that after 2011 or 2012, a commercial flight capability will be able to take astronauts and cargo to and from Earth to the ISS with their reusable craft as a replacement for the Space Shuttle.

The second private company originally named by NASA under the COTS program was Rocketplane-Kistler (RpK), headquartered in Oklahoma City. But towards the end of 2007, it was announced that they had failed to meet the stringent financial deadlines required under their contract. After a further review of the various alternatives, NASA announced in February 2008 that the second company to compete with Space X would now be the Virginia-based Orbital Sciences Corporation (OSC). Orbital is a long-established company, founded in 1982 by David Thompson, Bruce Ferguson and Scott Webster. In 1990, they successfully carried out eight space missions, highlighted by the initial launch of the Pegasus rocket, the world's first privately-developed space launch vehicle. In 2006 Orbital conducted its 500th mission since the company's founding. The COTS contract between NASA and OSC is worth $170-million and will be focused on the development of a launch system capable of delivering cargo to the international space station.

If neither Space X or OSC succeed in meeting the COTS objectives by the target dates, another option will be to use the NASA-developed Orion craft that is also supposed to fly astronauts to the Moon by 2015. An artist's conception of how this craft will look when first configured for low orbit flights is shown in Figure 1.9. NASA is also talking to Virgin Galactic and others about using their vehicles for both micro-gravity experiments and astronaut training. And the agency's interest in buying flights on the new vehicles also extends to using them for experiments in Earth's upper atmosphere as an analog for the thin atmosphere of Mars.

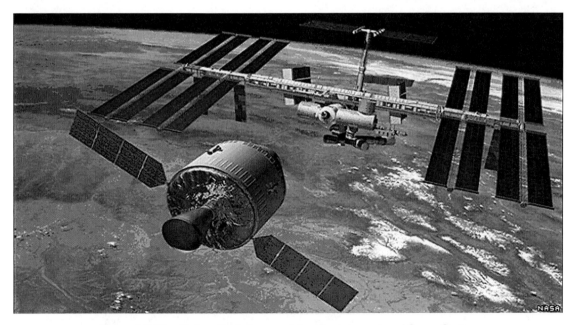

Figure 1.9: The Orion Craft Approaching the International Space Station
(Artist's Concept courtesy of NASA)

The Role of Prizes

Just as in the early days of aviation in the 1920's and 30's, the challenge of prizes for the development of space travel has proved to be a powerful stimulant in the development of new technology and new space capabilities. The new prizes are modeled after the Orteig Prize which offered $25,000 in 1919 for the first non-stop flight between New York and Paris. This prize was won by Charles Lindbergh in 1927 which changed the aviation industry forever.

The X-Prize, founded by Dr. Peter Diamandis and later sponsored by the Ansari family, touched off a ground swell of interest in developing the so-called space tourism market. It led to a range of efforts associated with both sub-orbital and orbital vehicles and June 2004 marked the start of what might be called the dawn of the "age of space tourism." This is how Mary Evans of *The Economist* described the event:

> "…space travel underwent its Wright-brothers moment with the first flight of *SpaceShipOne*. The roles of Orville and Wilbur were played by Burt Rutan, who designed the craft, and Mike Melville, who flew it – although they were ably assisted by Paul Allen, one of the founders of Microsoft, who paid for it. Of course, history never repeats itself exactly. Unlike the brothers Wright, who were heirs to a series of heroic failures when it came to powered heavier-than-air flight, Messrs. Rutan and Melville knew that manned space flight was possible. What they showed was that it is not just a game for governments. Private individuals can play, too."[9]

Figure 1.10: SpaceShipOne heads for the X Prize
(Copyright – Mojave Aerospace Ventures LLC, photo by Scaled Composites)

Since the success of the Scaled Composites team, who designed, built and flew the White Knight and SpaceShipOne vehicles to claim the Ansari X-prize, the enthusiasm to design and operate space vehicles to serve the space tourism market has continued to grow. Indeed, new projects have sprouted around the world and efforts are proceeding not only in the US but also in Argentina, Australia, Canada, Dubai, Europe, Israel, Russia, and the United Kingdom.

Prizes such as the Ansari X-Prize and NASA's COTS competition have now assumed an important role. In the Spring of 2006, a prize of $2-million was announced for NASA's Lunar Lander Analog Challenge. This prize is to go to a vehicle with the energy and accuracy needed to land on the surface of the moon. This design, because of the rough lunar terrain and the lack of landing strips, will obviously need a VTOL (vertical take-off and landing) vehicle that can navigate over the craters and debris found on the Moon's surface. The first competition for this prize was held in Las Cruces, New Mexico, in October, 2006 and the second in October 2007, but so far there have been no winners. John Carmack's vehicle, developed by another of the new players in the business, Armadillo Aerospace, however, came close.

This Moon lander competition event, under NASA's sponsorship, was held in tandem with another event organized by the dynamic Peter Diamandis (while still in his twenties, he was a co-founder of the International Space University and recently he was given the Arthur C. Clarke Innovator's Award). This new and still evolving event is called the X-Prize Cup. It is to be a series of "rocket races" with preliminary events held in Las Cruces, New Mexico. Diamandis envisions this competitive spectacle to be like a 21st century NASCAR race in the sky, with "rocket craft" navigating a three-dimensional computer specified racetrack – unique to each vehicle. This new and exotic sport has also been descibed as "Formula 1 racing on steroids." Peter Diamandis's stated goal is for these rocket craft ultimately to race right into space. In the meantime, this annual event, sponsored by the Rocket Racing League, is designed to attract fans and space enthusisasts from around the world and it is hoped that, if the analogy to Formula 1 races holds up, this new "sport" will become hugely profitable as well. So far the Federal Aviation Administration (FAA) has yet to establish definitive rules for such a potentially exciting but dangerous sport.

The "survival of the fittest" competition for successful sub-orbital designs to support space tourism will, in theory, carry over into "rocket racing" and the "Darwinian champion" from this natural competition should allow the best rocket technology to emerge as successful. It is ironic that some of the racing rockets and even some of sub-orbital spaceships now cost less than a Formula 1 racing car. Thus, this is a sport that appears likely to attract wealthy backers who see the market potential of this enterprise.

At this stage, the FAA has provided oversight and licensing for these events, but remains concerned about what type of safety controls and processes can best govern them in future. The FAA is trying to evolve a special set of rules to apply to such

Figure 1.11: Graphic of Envisioned Space Races in Las Cruces, New Mexico
(Artist's impression courtesy of X Prize Foundation)

an event that is in some ways like an "airshow" and in other ways a special new genre that needs special rules and regulations.[10]

Not only has NASA moved to encourage the development of new technology, but other major "prizes" continue to emerge as well. Perhaps the most demanding and certainly the largest space development prize is that created by the hotel magnate Robert Bigelow, who also founded Bigelow Aerospace in 2006. His announcement of the "America's Space Prize" (pictured right) supports the development of new technology for going into space for the longer term. The award amount is set at the very compelling level of $50-million.[11] Not surprisingly, the challenge to become the next Burt Rutan and claim a multi-million dollar prize is much greater than the Ansari X-Prize. The winner must accomplish a number of feats, be a US-based entity, and develop the winning spacecraft without government funding – although use of governmental test facilities is allowed.

The extremely difficult challenge involves building and flying a manned spacecraft that is capable of reaching an orbit of at least 400-kilometers altitude plus demonstrating the capability of carrying a crew of five and docking with a space station. The craft must then return safely to Earth and repeat the accomplishment again within 60 days. The spacecraft must complete at least two orbits of earth at the 400-kilometer altitude on both missions. The first flight must show that a crew of five can be carried, but the second one must actually carry a crew of five. Further, the spacecraft must essentially be reusable and very precisely no more than 20% of the craft can be expendable.

Perhaps most daunting of all, the deadline for claiming the prize is January 10, 2010. It is clear from analyzing the rules, or simply by talking to Robert Bigelow, that he is asking for a commercial space venture to develop an alternative capability to the Russian Soyuz craft and doing it within the next three years.

As the concept of the prize competition developed, Robert Bigelow tried to get NASA to co-sponsor this challenge to the emerging US private commercial space industry. But due to regulatory constraints and other reasons NASA did not participate. Bigelow is thus apparently considering purchasing insurance coverage to protect against the $50-million payout. At this stage, the demands of the competition are much more demanding than the X-Prize and the odds of successfully completing all of the requirements within the performance period seem small. Anyone seeking to become an official enrollee in the competition is required to register with Bigelow Aerospace and receive the detailed itemized rules at the Bigelow Aerospace web site.

In addition to offering the $50-million prize, Bigelow said his company is also is prepared to offer $200-million in conditional purchase agreements for six flights of a selected vehicle. He explained this concept in terms of his ultimate plan to deploy a private space station for space tourism and sees such a reliable private vehicle as critical to his mission. Bigelow has explained: "It could be somebody who doesn't win the competition, but who comes in late, but we like their architecture better than the winner's architecture."[12]

A New Prize from Google

In the commercial space development world, the excitement never seems to stop. The latest move came from the newest megacorp on the block, the Internet company Google. In the second half of 2007, they announced yet another new space prize to help stimulate the private space industry even further. This $30-million space prize is to go to the first company, university or organization to successfully land a rover vehicle on the moon's surface and broadcast images from the lunar surface. The true challenge is that this must be accomplished by a deadline of 2012 in order to collect the big bucks. The X Prize Foundation is also managing this competition and in February 2008 they announced an international field of ten entrants as listed in alphabetic order below:

Aeronautics and Romanian Cosmonautics Association (ARCA): The ARCA group is located in Valcea, Romania. ARCA, headed by Dumitru Popescu, earlier sought to compete for the Ansari X PRIZE. The craft they plan to enter in the Google Lunar X PRIZE is to be called the "European Lunar Explorer."

Astrobotic: Astrobotic is a hybrid organization that combines academic and aerospace industrial capability. This composite organization includes the Carnegie Mellon University, Raytheon Company several additional institutions. Carnegie Mellon's special is its stereovision for autonomous navigation. Astrobotic intends to compete for the prize using their "Artemis Lander" and "Red Rover" designs.

Chandah: Chandah, meaning "Moon" in Sanskrit. This new venture was founded by Adil Jafry, who heads the largest retail electricity provider in Texas. Chandah's spacecraft is to be rather exotically named "Sheherezade."

FREDNET: Is named with some immodesty after Fred J. Bourgeois III, who heads a multi-national team that is designed to combine various talents in systems and software development, space systems and propulsion.

LunaTrex: LunaTrex is comprised of several companies and universities. some of whom were also competitors for the Ansari X PRIZE. Each team member has special expertise in the various areas of rocket science, robotics, advanced energy systems, and propulsion systems. Their craft is to be named "Tumbleweed."

Micro-Space: Micro-Space, Inc. of Colorado has a 31-year history of producing high tech products. This company that is headed by Richard Speck has now flown 17 bi-propellant liquid fuel rockets as well as special designed rockets with vectored thrust guidance that can achieve "near hovering" capabilities. There engineers have also developed several innovative life support systems. Micro-Space has been a competitor in the Ansari X PRIZE and participated in the Northrop Grumman Lunar Lander Challenge. Their entry is simply known as "Human Lunar Lander" and is different from other entries in that it envisions the capability to carry humans aboard.

Odyssey Moon: Odyssey Moon is a private commercial lunar enterprise headquartered in the Isle of Man (UK) and headed by Dr. Robert Richards, a Canadian, and also one of the three founders of the International Space University. Odyssey Moon's craft is called "MoonOne (M-1)."

Quantum3: A US-based team, Quantum3 is seeking to develop a so-called launch-coast-burn trajectory that could ultimately achieve a soft landing on the surface of the Moon at the Sea of Tranquility. Quantum3 is a partnership of industry and university participants. Their craft is called the "Moondancer."

Southern California Selene Group: This group is based in Los Angeles area and headed by Dr. Harold Rosen, of the Boeing Corporation. He was at Hughes Aircraft Corporation when he designed the Early Bird satellite that initiated international satellite communications in the 1960. The craft, known as the "Spirit of Southern California" utilizes a simple design that is indicated to be inexpensive to build. The architecture for their "Spirit of Southern California" spacecraft will use control and communication systems first used in some of the earliest communications satellites (i.e. Early Bird) as well as the latest in electronic and sensor technology.

Team Italia: The final team in the initial group of ten is based in Italy and led by Prof. Amalia Ercoli-Finzi. Team Italia is a collaboration between several universities plus some industrial participation. The team is currently testing a prototype of its system at Politecnico di Milano.

It is understood that several additional registrants will be announced in coming months. The challenge for this contest is not only to complete all of the many complex tasks that have been devised, but to do so by the 2012 deadline.

The New Role of NASA

Time will tell if we have now entered a new era of space exploration where private investment plays an even greater role than before and governmental space agencies ultimately assume a secondary role. Certainly we appear to be in a period of transition, but it is clear that this evolution will take some time. Effective public and private cooperation will be needed to succeed. One of the key issues in this respect is whether the US government and NASA ultimately abandons, or largely abandons, all space efforts operating in low earth orbit and below to private enterprise.

The debate on NASA's future role was a feature of a panel discussion during a conference in Pasadena, California, in September 2007, where Burt Rutan insisted that "taxpayer-funded NASA should only fund research and not development." He said that the goal of private space tourism is to reduce the cost of space travel and exploration.

He added: "I think it's absurd they're doing Orion development at all. It should be done commercially," referring to NASA's lunar spacecraft.

NASA Administrator Michael Griffin responded to Rutan's vision in a speech at the same event. "Unlike Rutan, I will continue to think space programs are important," Griffin said. "We have here a program which is affordable, sustainable and which can be highly correlated to historical successes and developments from the past."[13] Many, however, would tend to quibble a bit with NASA's Griffin about how "affordable" the Orion and Ares development programs are. This is largely because the current Moon exploration program, that envisions a few "manned missions" to the Moon has a largish price-tag. It is, in fact, now projected to be well over $100-billion. As former Senator Dirksen once said: "A billion here, a billion there – after a while it adds up to real money."

Unlike the European, Japanese, Russian and Chinese Space Agencies, that continue to concentrate on space applications and missions that involve Earth-based activities, NASA is now focused heavily on The Cosmos – and especially the Moon, Mars and beyond.

Certainly, this much is true; private initiatives with private test pilots and private astronauts who are insured against risk (and their families provided for) add a new dimension of flexibility and risk management that government space agencies do not now have. One must hope that all of the new entrepreneurial space operators know – deep down inside – that failure means ruin for their programs while success and continued safe operation means world acclaim. Thus there is a clear and present danger that even one or two of the many organizations trying to develop systems to fly into space will give priority to being the first to fly commercial flights to space and overlook key safety features and concerns.

Private organizations such as Space X, Scaled Composites, SpaceDev, Bigelow Aerospace, etc. clearly have a different set of values and different types of incentive to succeed. The challenge for NASA, in its role of space explorer and developer of new launch systems, is to find new and flexible ways to fuel new intellectual energies, find better technology, and create safer space systems and yet do this through "conventional governmental" processes. The incentives that the FAA, NASA and Congress have created to stimulate the space tourism business, plus the magnetic allure of the Ansari X-Prize, all clearly seem to be working. It is these powerful stimulants that have triggered a large number of established and start-up companies to design launch and space plane systems to achieve sub-orbital or even low Earth orbit access.

There are people at NASA, such as astronaut Neil Woodward, that are managing the COTS commercial access to space program, who see a greatly expanded commercial role in human space flight to the space station and beyond. Others, such as Kenneth Davidian, are also seeking within NASA a new "commercial development" policy for future space programs. The Aldridge Commission report, that was largely ignored during NASA Administrator Sean O'Keefe's stay at NASA's helm, is being viewed in a new light at least by some. This new vision for NASA, however, is still a nebulous concept for many.

Successes – and early setbacks.

The development of spaceplane technology and the desire to provide an opportunity for "space tourists" to fly into low earth orbit is not simply a matter of R&D and engineering. The business and market aspects are perhaps equally important for all those seeking 'a license to orbit'.

This 21st century phenomenon in many ways parallels the early days of the aircraft industry. Just as in the early part of the 20th century there was a the plethora of planes being designed and built, today's world is beset by a diverse group of companies, large, medium and small, that are designing craft that they hope will take hundreds and then thousands of space tourists into the black sky of sub-orbital space. Others aspire to build systems that can go into orbit and even allow space tourists to walk in space or stay at a space hotel, as envisioned by Arthur C. Clarke in *"2001: A Space Odyssey"*.

These various efforts largely share two goals that in many ways are in conflict. The space entrepreneurs want to be first to launch space tourists into this wondrous and largely unknown environment. Yet there is recognition that a space tourism enterprise will only succeed if the flights are safe and passengers return safely from these excursions.

But just as in the early days of flight, and in the first stages of the US and Soviet space programs, the development of private space has not been free of setbacks. During 2007, Burt Rutan's Scaled Composites company suffered a serious explosion during the testing of a rocket engine for the SpaceShipTwo vehicle at their facility in Mojave, California. The explosion resulted in the death of three workers and injuries to others and this commercial space tragedy has been compared, in its seriousness and impact, to the Apollo 1 accident in 1967 that led to the explosive deaths of three astronauts on the Saturn V launch pad.

The Mojave accident, for which the US agency OSHA has fined Scaled Composites for safety failings, underlined that this is still an experimental and dangerous undertaking. This mishap has, on one hand, certainly led to a delay in the original timetable for commercial sub-orbital flights; but it has perhaps helped to make sure that the commercial launches that do occur will be better engineered and hopefully safer.

Nor was the "tragedy in the Mojave" the only set back for commercial space development. The first three unmanned Falcon rockets launched by the Space X company failed during their test flights. But, Space X's Elon Musk, with true entrepreneurial spirit, made it clear that much had been learned from these failures and in September 2008, his Falcon 1 became the first privately-developed rocket to orbit the earth.

However, these costly disasters are clear reminders that space flight is a hazardous business and that there will inevitably be difficulties to surmount along the way. After the Apollo 1 accident in January 1967, NASA did a complete review of its rocket design and test program from stem to stern. Perhaps "propelled" by this incident, NASA recovered and memorably succeeded in landing the first astronauts on the moon just 29 months later in July 1969. Such setbacks, even human tragedies, lead to rethinking, redesigning and caution. We can only hope that this means that the space tourism business is "working out the kinks" before real passengers fly.

Nearly 50 New Companies

The field of commercial space companies is increasingly international and crowded. There are now nearly 50 such enterprises involved in developing space planes, lighter-than-air craft that can fly into orbit and spaceports. Most of these firms are in the US, but a number are in other countries around the world. All of them, big and small, are playing a part in developing the space tourism business. They range from "service companies" such as Space Adventures, that can arrange dozens of alternatives for would-be space travelers, to research and technology companies working on new ways to provide access to a sub-orbital space ride to low earth orbit or even beyond. Appendix A at the end of the book summarizes these entities, the means by which they are seeking to provide access to space, the names of their craft and their current goals and objectives. In many cases these companies envision a staged development with expanded capabilities over time.

Space tourism is clearly a highly fluid "industry" reminiscent of the barnstorming days of early aviation a hundred years ago. A great deal of shake-out will continue to occur with some dropping out, going bankrupt, or merging with others. Already over two dozen hopeful commercial space companies have bitten the dust. Meanwhile, other new entities or subsidiaries of established firms will be joining the process of developing new, reliable and affordable access to space. Maintaining safety, setting reasonable and viable risk management rules and regulation, and undertaking effective independent verification and validation processes for this totally new type of industry is clearly a great challenge. This will require a great deal of balance and judgment. On one hand there must be the opportunity for innovation and totally new approaches to space travel to obtain new and better solutions, yet at the same time there must be a stringency of safety controls and inspections to prevent needless and foolish loss of life. In the US, the FAA is designated by the Congress to be in charge of overseeing this burgeoning new industry. Other countries are also beginning to consider how such activity is to be regulated and they are looking very closely at the Rule Making that the FAA has established in response to the amendments to the Commercial Space Launch Act of 2004.

Many centuries ago, or so the story goes, a would-be astronaut in China sat atop a cluster of rockets powered by gunpowder. Thus he sought ascent to the heavens. But instead, to the horror of the assembled witnesses, he was rather immediately incinerated.

Fortunately we have come a long way in rocketry and space travel since that time. Careful and prudent safety regulation of this newly emerging "industry" is essential to protect the foolishly innovative and overly daring inventor and entrepreneur. Care is also needed to protect the property and lives of others. And perhaps equally important to the aerospace world is the need for safety regulation and thoughtful licensing activities in order to prevent rash and risky experiments that spoil or retard the development of an entirely new industry and type of human endeavor. In short, the risk is not only to the flight crew and passengers of a dangerous vehicle; rash and premature efforts to "jump start" the space tourism business, may destroy the initiatives of many who have well-conceived ideas as to how to provide safe access to space.

But above all, we are seeing a tremendous surge of enthusiasm and creativity. Engineers, innovators and entrepreneurs abound. They all seem to think that they can develop new and better space plane technology. Many will fail, but it will only take a handful of the very best to support the growth of a new space tourism industry.

Although the prospect of space tourism seems scary and to some almost science fiction, the challenge of sub-orbital flight is much less than actually going into space. The following chart (Figure 1.12) shows the relative speeds and thrust levels need to travel by airplane, jet, space plane and rocket to outer space. Space planes, in terms of energy requirements can thus be clearly seen to be more like a jet plane ride than going into Earth orbit.

The range of space tourism activities will, if anything, likely continue to expand as the business matures. At the low end will be Earth-based simulations of space flight, while organizations like Space Adventures will continue to push the envelope for truly wealthy space explorers seeking to accomplish new firsts. The next space tourism frontier apparently will be a $100-million ride around the Moon on a Soyuz vehicle, with perhaps civilian space walks also planned for the future. Science fiction just keeps getting more and more difficult to write.

Comparing Airplanes, Jets, Sub-Orbital Space Planes and Rockets to Orbit				
Comparative Factor	Airplane	Jet	Spaceplane	Rocket to LEO
Velocity (m/sec)	250	500	1600	7800
Height (km)	10	20	100	200+
Specific Energy (Joules/kg)	0.13	0.7	14.5	324
Energy Ratio	0.0004	0.0022	0.044	1

Figure 1.12: Putting Spaceplane Travel in Perspective
(Courtesy of data provided by Prof. Nikolai Tolyarenko, International Space University, 2007)

References:

1. The Statesman, London, 27 April 2007: http://www.thestatesman.net/page.news.php?clid
2. Dennis Tito testimony –
   http://www.spacefuture.com/archive/hearing_on_space_tourism_testimony_by_dennis_tito
3. Space Future Journal – http://www.spacefuture.com/journal/journal.cgi?art=2006.11.01
4. Space Adventures – Sept. 28, 2006 – Skynews.ca (http://www.spaceref.ca/news/viewpr.html?pid=20925)
5. abs-cbnNEWS.com/Newsbreak.
6. PopularScience.com: "Space Adventures Charters Entire Russian Spacecraft" – July 2008.
7. Space Adventures – http://www.spaceadventures.com/index.
8. "Riches to Ride: Local Firm Launches Wealth Thrill-Seekers Into Space" Washington Post, April 16, 2007, p. D 1-2.
9. Mary Evans, "Rocket Renaissance: The era of private space flight is about to dawn" – The Economist, May 11, 2006 http://www.economist.com/science/displayStory.cfm?story_id=6911220)
10. R. Repcheck, "Unique Public Safety Issues Associated with Rocket Shows" – Proceedings of the Second Annual Conference of the International Association for the Advancement of Space Safety, May 14-16, 2007, Chicago, Illinois
11. Leonard David, "Rules Set for $50 million "America's Space Prize" – Space News
    www.space.com/spacenews/businessmonday_bigelow_041108.html
12. Leonard David, "Rules Set for $50 million "America's Space Prize" – Space News
    www.space.com/spacenews/businessmonday_bigelow_041108.html
13. ZDNet Technology News – ZDNet News: Sept. 20, 2007

— Chapter 2 —

"The past few decades have been a dark age for development of a new human space transportation system. One multi-billion dollar Government program after another has failed ... The public, whose hard earned money has gone to fund these developments, has felt it indirectly. ... When America landed on the Moon, I believe we made a promise and gave people a dream. It seemed then that, given the normal course of technological evolution, someone who was not a billionaire, not an astronaut made of "The Right Stuff", but just a normal person, might one day see Earth from space. That dream is nothing but broken disappointment today. If we do not now take action different from the past, it will remain that way." – Elon Musk in evidence to the US Senate Hearing on Space Shuttle and the Future of Space Launch Vehicles in April 2004.

THE BILLIONAIRES ENTER THE COMMERCIAL SPACE RACE

If you want to start a new industry, there are lots of good ways to begin. A totally new idea such as the Internet, a sudden new consumer market such as plastics, or great capital financing opportunities, such as an eager new IPO market – these are all good beginnings. The space tourism industry, however, has attracted an extra-special ingredient. Many of the new "space tourism entrepreneurs" are self-made multi-billionaires. The mainstays of this new business are very successful business people, all with a touch of exotic glamour and press coverage, plus gigabucks at their command.

Most billionaires become wealthy by innovative thought and shrewd investment. The space tourism leaders are certainly no exception to the rule. This amazing line-up of space tourism nabobs includes Virgin Atlantic head Sir Richard Branson, Amazon.com founder Jeff Bezos, Pay Pal wizard Elon Musk, Microsoft co-founder Paul Allen, video game designer John Carmack, (who developed both "Doom" and "Quake"), hotel magnate Robert Bigelow of Budget Suites fame, and the Iranian-Dubai-American Ansari family, whose wealth springs from a number of sources. Just to add some additional spice, there are also some old-fashioned mere multi-millionaires such as Jim Benson, who made his first fortune in the computer industry, and Eric Anderson, who has already made a handsome sum in the space tourism business that he largely created. Plus there are others still emerging. These include David Thompson's successful Orbital Sciences Corporation, which has won the re-bid contract from NASA for the Commercial Orbital Transportation System (COTS) development. Thus Elon Musk's Space X and Thompson's Orbital Science are now going toe-to-toe to develop the first truly commercial vehicle that could take cargo and ultimately people to orbit and back.

Virtually all of these entrepreneurial drivers of the new space tourism industry share a common trait. This is the dogged determination to find new and innovative ways to access space – plus an inclination to discard outmoded concepts and conventional ways of thinking. These are the people who, like Alexander the Great, tend to pull out a sword and hack through the Gordian knot rather than taking the slow and conventional approach to problem solving. They are not interested in bureaucratic red tape, and as a rule, they throw out the window the cumbersome processes found within traditional governmental space agencies. This can cause some reasonable jitters when it comes to space safety concerns. The explosion in the Mojave Desert in 2007, when the SpaceShipTwo rocket motor blew, up renewed these concerns.

This new breed of space plane developers believes they can accomplish their mission by hiring dozens of the most talented and dedicated employees and focusing on a very clear and precise mission – to enable the masses to fly into space. It may be proto-space and the flight above 100-kilometers might last only about four minutes, but it is still high enough to see the blue ball of the Earth in the sky against the darkness of space. For many starry-eyed citizen astronauts this is more than enough.

This approach stands in contrast to creating and staffing huge national space agencies and large aerospace corporations that command vast complexes and labs with thousands of employees. These old guard space enterprises look like vast, high tech armies from the military-industrial complex; by contrast, the new entrepreneurial organizations with dozens or perhaps scores of employees look, act and feel very different from the space agencies in their approach. These new and sometimes brash organizations put their "can do" bravado out front.

The New Commercial Space Industries as the "Anti-NASA"

The "lean and mean" entrepreneurial space companies that are being created by these billionaire business people are, if you will, in mood and temperament the "anti-NASA." This is to say their approach is antithetical to the typical NASA approach to space-related activities. They are thinking differently and their approach is innovative and unconventional in terms of organization, staffing and goals. The starting point for most projects is "outside-the-box" thinking. These new organizations are seeking to provide mass public access to space that is low in cost, safe, hassle-free, AND achievable in the near term.

In this quest, they are looking for ways to extend the technology and reliability of the aviation industry and to upgrade it to achieve sub-orbital flight. They are trying to extend proven technology another notch up, rather than thinking of ways to develop experimental rockets to go to the Moon. They do not see their mission as the development of difficult and expensive state-of-the-art technology or totally new materials. This mission is seen as being better left to governmental agencies and aerospace giants. They only ask the government to stay out of their way. Confident? Yes! Brazen? You had better believe it! Irresponsible and hopelessly cocksure? Time will tell.

Their "business models" are based on the idea that they want to reduce costs and develop affordable transportation systems. They also recognize that they have to convince potential space tourists that not only will they fly into space, but that they will also come down – safely. As the billionaire adventurer and businessman Sir Richard Branson, now head of Virgin Galactic, has said: "I'm absolutely sure that millions of people want to go into space and its up to us to make it affordable."[1]

The first round of a hundred "Founders" are already booked to go on the first flights of Virgin Galactic's SpaceShipTwo. The seats are fully sold out at $200,000 each and some of them, including Branson himself and his son Sam started their training in January 2008 at the National Aerospace Training and Research Center in Bucks County, Pa., USA. Branson has also said that the first passengers will include his 92-year old father and 89-year old mother. Sir Richard knows how to hype the safety of his newest enterprise, even though the FAA waiver statement notes up front that space tourism flights are a "high risk" venture.

We have a long way to go to achieve the "still to be met" space safety goal, set over two decades ago. This goal was to experience only one space fatality per one thousand space flights. The reality is that the national space agencies have had fatalities resulting from 1% of all flights with humans aboard. In fact of the 500 or so astronauts and cosmonauts that have flown, 22 have died. This is 4% of those who have made it to space and back. (The difference between the 4% fatalities for those that have flown and the 1% mortality per flight is that many have flown on multiple missions.)

The new billionaire space entrepreneurs see the prospect of not only developing new approaches to space flight, but also ways to embark on profitable space enterprises based on mass market volumes. These initiatives require an entirely different business model from that pursued by a NASA or the other space agencies around the world. The starting point has been to find economical ways to exploit the safest jet and rocket propulsion, using conventional vehicles to reach "spy plane" altitudes, proven rocket technology, and modern and highly cost-effective manufacturing and testing facilities. In short, the space agencies think in terms of multi-billion dollar programs lasting decades; whereas the new commercial space people think in terms of million-dollar projects with turn around schedules of 3 to 4 years.

There is a famous concept attributed to Thomas Ockham called Ockham's Razor: "If there are two solutions, pick the simplest." This advice seems to be the constant guide to the emerging space tourism business.

Space plane designers are constantly seeking new ways to ensure passenger and crew safety, including reasonable and effective "lifeboat" strategies. The space plane technology involved is more akin to that of a supersonic experimental jet than that of a moon rocket. The speeds involved are more like Mach 4 or so, rather than orbital or escape velocities that are many times higher. Table 2.1 below notes just how different the space plane technologies are from Moon missions.

And all this should be seen as just the start of a new type of commercial activity into space based on the "Anti-NASA" model. The hope is that it will lead to new ways of looking at space as an entrepreneurial

Difference in Characteristics	Reusable Orbital Launch System	Sub-Orbital Spaceplanes
Maximum Velocities	Up to Mach 30	Mach 4 to 6
G forces	Very High G forces	3 to 5 g (during descent)
Thermal Gradients on Reentry	Thousands of Degrees C	Hundreds of Degrees C
Environmental Protection Systems and Structural Strength of Vehicle	Very demanding in terms of design and materials	Much less demands in terms of structural strength, atmospheric systems, life support, etc.
Exposure to Radiation	Can be high levels	Minimal exposure due to short flight duration and lower altitudes
Exposure to Potential Orbital Debris Collisions	Exposure increases as length of mission increases	Exposure risk is very low due to short duration and lower altitudes
Escape Systems	Parts of the flight during high thermal gradients make escape systems extremely difficult and expensive to design.	Escape systems are much easier to design due to lower thermal gradients, lower altitude, etc.
Type of flight suits required	Expensive and complex flight suits required	Simple and lower cost flight suits are required due to lower altitudes, lower thermal gradients, much shorter exposure to low oxygen atmosphere.
Launch Risk Factors (overall)	Very high	Considerably lower and different

Figure 2.1: Comparing Orbital Launch Systems (Reusable) with Sub-Orbital Spaceplane Systems

business across a wide spectrum of opportunity. Most of the billionaires and entrepreneurs involved see this ultimately as much more than just offering pleasure trips up into the dark sky above the Earth's atmosphere. The ultimate frontier and the new "space business plans" need to be more than joy rides.

Possibilities Beyond Space Tourism
Over time, it is hoped that these new commercial space ventures may expand into other vital areas. These might include the deployment of solar-powered satellite systems to provide clean, low carbon-polluting energy at affordable cost. This may need to be part of a greater eco-strategy to save the world from runaway global warming.

The future might also see the deployment of a space elevator that reaches beyond the geosynchronous orbit (or Clarke orbit – in honor of Arthur C. Clarke, who first identified it in 1945). In time, this cable to the sky could possibly allow communications or other types of satellites to be placed into orbit at very low cost – perhaps $10 a kilogram rather than the $10,000 to $20,000 per kilogram required of today's inefficient and polluting chemical rockets. These new space entrepreneurs may allow new markets to develop to support low cost material processing in near- zero gravity environments.

These possibilities are only beginning to be explored by business people who see the new commercial space frontier as an economic opportunity rather than a research laboratory. Who knows, within a decade or two or three, we might be able to fly to spectacular space hotels or journey to space-based laboratories where new and better drugs can be manufactured to combat cancer or other diseases. Indeed there may be dozens of new commercial objectives yet to be identified. The national space agencies have already spun off scores of commercial applications over the past decades of the space era – and with a lot more on offer than Tang and Teflon. The space entrepreneurs, with their business acumen and marketing expertise, may not only identify many more commercial applications, but exploit them sooner than governmental agencies would. We can at least have high hopes for the space exploits of successful men like Paul Allen (co-founder of Microsoft) who financed SpaceShipOne, or Elon Musk, Jeff Bezos and Richard Branson. Such innovative and fearless thinkers can reasonably be expected to be more adept at bringing these new products or services to market sooner and with greater flair.

The key to focus on here is that the billionaire space entrepreneurs can make a difference. At the same time, there are also other space plane pioneers, who may be without the big bucks and fame of their billionaire colleagues, but who still have lots of determination. The dozens of other smaller-scale space tourism developers working on space plane systems have also dared to envision totally different business models for their enterprises – often on shoestring budgets. Their approach too has been much different from that of a NASA or other space agencies. They have looked at market demand, service needs and product development as business innovators, and not as governmental research scientists. They see space tourism as not just a scientific and technical mission, but rather as a business that will first be based on providing entertainment or a "vacation experience." This is admittedly a big step beyond hot air balloons or bungee jumping. But this innovative thinking has led to innovative designs involving lighter-than-air craft, ion engines, hybrid fuels, combining high altitude spy plane technologies with rocket systems and other unconventional approaches.

Many see the prime way forward as a "tourist experience and entertainment enterprise." This vision would stress that all offerings to the mass public market must be safe, secure, convenient, and affordable – and yet still be exciting and awe inspiring. Such a business model requires eliminating "hassle, inconvenience, or fear" from the space tourism experience while making passengers feel they have undertaken a once-in-a-lifetime journey that sets them apart from mere Earth-bound dwellers. Of course, after the training and the briefings, and after the corporate and the FAA waivers are signed to acknowledge they are undertaking a death-defying feat, space tourists may still feel a bit less than safe – unless they are entirely clueless.

But the space entrepreneurs – large and small – are certainly working behind the scenes to ensure a maximum of safety, competence, and emergency escape processes while also streamlining operations and developing systems that can reduce cost. In the US, the Federal Aviation Administration Office of Space Commercialization (FAA-AST for short) has undertaken an elaborate rule-making process. They have now adopted regulations to govern space tourism flights, the pilots, the crew and launch range safety officers. In this process they have tried very hard to accommodate to these new business models and respond favorably to the comments received from the various members of the newly formed Personal Spaceflight Federation (PSF). With a host of billionaires within its membership, this Federation is one of the most exclusive "clubs" in the world.

The personal spaceflight industry is not just a US enterprise. Although three quarters of the space tourism and space plane businesses are US-based, quite serious efforts are also underway in Canada, China, Israel, Russia, the United Kingdom, France, Germany and Switzerland, among others. In some countries, such as China, France, Germany, India, Japan and Switzerland, the emphasis on development in space tourism and space planes has been often been in the public sector as opposed to truly commercial enterprises. In the case of Russia, Roskosmos (the Russian Federated Space Agency) has indicated a willingness to work in parallel with commercial aerospace concerns while maintaining a governmental program as well.

Although most of the first commercial space tourism businesses will be US-owned, some, like Virgin Galactic, will be U.K.-owned but operated from New Mexico. Others such as Space Adventures, may operate from spaceports in the United Arab Emirates, Singapore or Malaysia, but still be US-owned enterprises. Even so, the rest of the world, in terms of trade considerations, national security concerns, and international aviation regulation, will necessarily be involved. At this point, the rest of the world seems willing to follow US regulations until they are proven wrong.

The British Knight
Among the remarkable group of entrepreneurial billionaires the most headlines to date have been grabbed by the U.K.'s Sir Richard Branson. He is the odds-on favorite to become the first to initiate "space tourism" with his Virgin Galactic flights on the SpaceShipTwo craft. By becoming an investor in the SpaceShip Corporation, following the X-Prize success of Burt Rutan and Paul Allen (Rutan built and Allen financed SpaceShipOne), Branson has managed to become both a space plane manufacturer as well as a thrill marketeer.

Branson has invested in the second generation of a proven concept and he plans to have the first of his new spacecraft flying tests in 2009, with the "Founders" group of "tourists" going into "near space" in 2010. Some $15-million in reservation fees has already been collected for future flights and the list of prospective clients is impressive, including movie stars, sports figures and a royal and two bookings involving marriage, one to get married in space and the other for the couple to have their honeymoon in space.

Now Professor Stephen Hawking, the world's leading astrophysicist and cosmologist, who recently completed his first Zero-G flight in spite of his afflictions, has said he plans to take a space flight as well.

Figure 2.2: Sir Richard Branson shows off SpaceShipTwo
(Photo courtesy of Virgin Galactic).

Branson's views are not necessarily consistent. He talks about his newest venture as part business, part amusement park ride, and part visionary mission to save the Earth. This is revealed in his various statements about developing rockets for supersonic transport between New York and London or perhaps colonizing the "Virgin Moon" and then on to global ecology: "I hope that people going into space will come back and appreciate this beautiful world more," he said.[2]

Branson, not one to do things by half measures, ordered not one but a fleet of five SpaceShipTwo vehicles at a cost exceeding a quarter billion dollars. This fleet is expected to fly his "space tourists" on sub-orbital flights to an altitude of over 100-kilometers so they can experience weightlessness, dark sky and see the Earth as a large blue marble in space. (Actually, it should be pointed out, in terms of truth in marketing, the "payoff" period of the flight with weightlessness "in outer space" will likely only be about four minutes long.) Under the deal being negotiated with Rutan and Allen, Virgin will pay up to $21.5-million for an exclusive license to SpaceShipTwo's core design and technologies. Another $50-million will go to Rutan's company Scaled Composites to build the five passenger spaceships. An equal amount will be invested in operations, including the Virgin earth-base in the desert.

According to Will Whitehorn, President of Virgin Galactic, the company has already banked $38 million in deposits. "We always said it was going to cost us $250 million to get this off the ground, now it's looking like $300 million," he said in July 2008. Even so the business plan is predicated on breaking even in 2012, and making profit thereafter." Whitehorn admits that this is a high-end spend, but says there are enough customers out there who want to take part to make the venture a success.[3]

The current business plan envisions 50 passengers a month, paying $200,000 each for a two-hour flight to an apex beyond Earth's atmosphere, wrapped in a three-day astronaut experience. $10-million a month seems like good income, but against considerable expenses and risk management fees, it may be a marginal business plan, especially if there should be a mishap that closes down operations for a significant period.

Like the X-prize winning design, Burt Rutan's Virgin Galactic spaceships will use the 'piggy-back' systems with the new SpaceShipTwo carried to a height of about 60,000-ft by the White Knight Two. In July 2008, just a year after the fatal accident at the Mojave CA location, the first White Knight Two was ceremonially

Figure 2.3: The real thing" – Virgin Galactic's White Knight mothership
on its test flight program over the Arizona desert in early 2009.
(Photo courtesy of Virgin Galactic)

rolled out from its hangar for the first time. This first all carbon composite two-skinned aircraft has a 42m (or 140-ft wingspan) and was 'christened' Eve by Sir Richard Branson for his mother, who is expected to fly on the first passenger flight. Test flights for White Knight Two are now underway in the skies over Arizona and the first passenger flights into space and back are planned for 2010.

Virgin Galactic reports that in addition to the 250 individuals signed up to be among the first to fly aboard the history making service, they also have around 85,000 registrations of interest.[4] This represents a deposit base exceeding $35-million and more than $45-million of future income to the fledgling spaceline. By March 2008, eighty of SpaceShipTwo's first passengers had already been through medical assessment and centrifuge training at the NASTAR facility in Philadelphia.

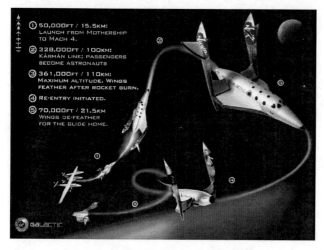

① 50,000FT / 15.5KM:
LAUNCH FROM MOTHERSHIP
TO MACH 4.

② 328,000FT / 100KM:
KÁRMÁN LINE; PASSENGERS
BECOME ASTRONAUTS

③ 361,000FT / 110KM:
MAXIMUM ALTITUDE. WINGS
FEATHER AFTER ROCKET BURN.

④ RE-ENTRY INITIATED.

⑤ 70,000FT / 21.5KM
WINGS DE-FEATHER
FOR THE GLIDE HOME.

Figure 2.4: A ride on SpaceShipTwo
(Diagram Courtesy of Virgin Galactic)

Ironically, one of Branson's largest problems has been that the US government's ITAR (International Trade in Arms Regulations) processes have slowed the transfer of technology to the U.K.-based Virgin Galactic. However, the plans are moving ahead and when they start to operate, the SpaceShipTwo vehicles will fly almost vertically until the craft reaches maximum velocity. The passengers and crew will then coast upwards into a parabolic arc where weightlessness will be achieved for about 4 minutes. At this time, passengers will be able to enjoy the spectacular view of the Earth below. On the descent, rocket engines will be fired to pull out of the nose-dive and the craft will ultimately land at the spaceport from which it took off. The entire experience will be about the time of an airline flight from Washington, D.C. to Atlanta, Georgia. In a nutshell, it will be more an experimental test pilot trying out a very high altitude jet than an astronaut going to the Moon or even the International Space Station. A whole lot of sensations will be packed into a very condensed period of time. If it was taking place in an amusement park, a teenager would be inclined to say: "Let's do that again." The problem is that the price tag will prevent many people from signing on for the next available flight.

The high price means that a lot of value-added will be provided on the ground in terms of training and preparation so that passengers will better appreciate the worth of their $200,000 or so investment. If anyone can pull this off as a successful business Sir Richard Branson seems the one most likely to do it.

The Richard Branson story

The multi-faceted Richard Branson has had a dynamic and widely-publicized business career. He was born in 1950, in Surrey, U.K. and is now Britain's most famous entrepreneur, best known for his Virgin brand name that now encompasses a wide variety of businesses. It is estimated that Branson is worth over $6-billion, according to the Sunday Times "Rich List of 2007." He is described as a 'transformational leader', with his maverick strategies for the Virgin Group as an organization driven on informality and information, and not strangled by top-level management.

An entrepreneur from an early age, he is reputed to have started two failed ventures by the age of 15 – a Christmas tree growing business and a pigeon-raising farm. Surprisingly, he was not a very good student, suffering from dyslexia. Richard was a good sportsman at school, but it was a serious athletic injury that helped to launch his illustrious career. At 16, he decided to quit school and moved to London where he began his first successful entrepreneurial activity. He identified the energy of student activism in the late 60's and started his own newspaper called "Student Magazine". It was focused on the students, not the schools, sold advertising to big business and included feature articles by Government ministers, rock stars and movie celebrities. The headmaster of his school wrote: "Congratulations, Branson. I predict that you will either go to prison or become a millionaire."[5]

Then in 1970, he set up a record mail order business and two years later opened a record shop in Oxford Street, London. Next, he launched the record label Virgin Records, and opened a recording studio. Apparently, the name "Virgin" came about when one member of his group said: "we're complete virgins at business."

The company's first disc was multi-instrumentalist Mike Oldfield's *Tubular Bells*, which was to become a best seller. In fact, Virgin secured the album because no other company was prepared to produce such an unconventional record. Branson's company also courted controversy by signing bands like the Sex Pistols and won praise with obscure avant-garde music such as the so-called "krautrock" bands "Faust" and "Can." Later, in 1992, to keep his airline company afloat, Branson sold the Virgin label to EMI.[6]

Following his success in the music industry, he created Virgin Atlantic in 1984, offering reasonable fares on scheduled transatlantic flights. It grew to become Britain's second-largest carrier and once again, he had spotted a gap in the market and filled it spectacularly. In 1997 he took what many saw as one of his riskier business exploits by entering the railway business. Virgin Trains won the franchises for two sectors of the former British Rail network. This venture promised new high-tech tilting trains and enhanced levels of service, but this proved hard to deliver. Aging trains, tracks and facilities were hard to replace in short order and Virgin Trains faced problems in the late 1990s as they sought to make trains run on time while it awaited the modernization of track systems and the arrival of new rolling stock.

In 1999, he moved into telecommunications by launching Virgin Mobile, followed by Virgin Blue in Australia in 2000. Then, in 2006, he combined Virgin Mobile with a leading UK cable TV company, NTL. This new combined company was renamed Virgin Media.[7]

Branson has also outlined plans to acquire Britain's commercial TV network. This would create an $18-billion television giant that would seek to compete with Sky television and its dominance in pay TV in the UK market. Sky then stunned the media world when it paid nearly $2 billion for a 17.9% stake in ITV, a move that was widely perceived as an attempt to block a potential tie-up between Virgin Media and ITV. Branson immediately claimed that Rupert Murdoch's Sky was trying to distort competition in the TV market and called on regulators to take action. The outcome of this clash of business titans between Rupert Murdoch and Richard Branson is still to be determined.

Among many other enterprises, Branson entered the financial services market in the UK with Virgin Money, and tried unsuccessfully to take over a failed bank, Northern Rock. He developed a Virgin Cola brand, and even Virgin Vodka (which has so far not been an overly successful enterprise.) In short Branson, who is an avid explorer, balloonist and risk taker par excellence, seems to have no limits to his diverse interests and commercial ventures. In 1999, he received a knighthood from the Queen for his services "to British entrepreneurship." He is married and lives with his wife Joan and their two children, Holly and Sam, in London and Oxfordshire, UK – and on his own Caribbean island.

Microsoft Co-Founder Paul Allen and SpaceShipOne

But Branson is not alone in a club of billionaires with the itch to soar into the stratosphere via a space tourism business. Paul Allen, the Microsoft business entrepreneur, has already made his mark in the world of space by claiming the Ansari X-Prize together with Burt Rutan. He was the money behind the spectacular performance of SpaceShipOne.

Paul G. Allen hails from Seattle, Washington where he was born in January, 1953. He went on to co-found one of the world's most lucrative and influential companies, the Microsoft Corporation, in 1974. The company's products revolutionized personal computing and made billionaires of Paul Allen and Bill Gates. Allen actually left the company in 1983 due to illness, and has since invested in a wide variety of projects covering the technology, entertainment, sports, and aerospace fields. He reportedly invested as much as $100-million in backing Rutan's team in their bid to win the X-Prize. Although this may be merely chump change for NASA, it was serious money for a private entrepreneur. But, they prevailed. Burt Rutan and Paul Allen became the winning combination that not only won the X-Prize but they went on to join with Richard Branson to create the new SpaceShip Corporation. This company is now playing a key role in this new industry, starting by supplying the initial fleet of space planes to Virgin Galactic.

Figure 2.5: Paul Allen gets a Progress Report
from Burt Rutan
(Photo courtesy of Scaled Composites)

But Allen and Rutan certainly do not expect to stop there. Eventually, they hope to supply some forty SpaceShipTwo craft to the global market. Some may scoff, but Rutan has made a habit of out of doing the impossible. He designed the Voyager aircraft, whose long and elegant wings were also high capacity fuel systems enabling it to fly around the world non-stop without refueling. He has also designed various Unmanned Autonomous Vehicles (UAVs), military aircraft capable of flying long duration missions. All of these strange craft, a number of them "classified" in nature, emerged from his Scaled Composites facility in the Mojave Desert. It was one of Rutan's experimental planes that the singer John Denver was flying when he tragically crashed and died in October 1997.

Of the dozen and a half entries for the Ansari X-Prize contest, the Paul Allen and Burt Rutan company, Scaled Composites, was considered to be one of the few teams that might succeed. Rutan had already achieved almost legendary status as a designer of new types of aircraft and he successfully met the challenge of flying crew and passengers above 100-km and repeating the flight within eight days. His innovative design involved a two-tiered approach, by flying the so-called White Knight "carrier vehicle" up into the stratosphere and then releasing SpaceShipOne to fly into "outer space."

One of the keys to safe re-entry of the SpaceShipOne vehicle was to achieve a relatively slow descent that did not involve high temperature gradients. Rutan diagnosed high temperatures as a key safety problem and instead of developing a thermal protection system to protect against dangerous high and ablative heat (as NASA and other space agencies have always done) he concentrated on getting the heat levels down. SpaceShipOne also used an innovative engine that employed a hybrid fuel system with neoprene rubber as the fuel and nitrous oxide (i.e. laughing gas) as the oxidizer. This system was developed by the SpaceDev Corporation and it allowed the solid fuel system to be the "throttled" so that if a problem developed the laughing gas would shut down and so would the engine. There are always pros and cons and pluses and minuses when trying to design unique and break through systems. The nitrous oxide and neoprene fuel gives off a residual exhaust that is much "dirtier" than many other rocket systems.

In August 2008, the relationship between SpaceDev and Scaled Composites was renewed. The new multi-year contract specifies that SpaceDev will be the lead developer on the rocket motor team for the SpaceShipTwo vehicle. This means they will collaborate with Scaled's internal design team to develop a production-ready hybrid rocket motor. SpaceDev will conduct ground tests on the motor components and assist Scaled in the full-scale rocket test program both on the ground and during SpaceShipTwo flight tests. The contract, which runs through 2012, has an initial value of approximately $15-million.

Ground Test Fatalities

Although the program seems to be on track today, in July 2007 the program seemed to crash and burn – literally. This was when there was a fatal accident at Scaled Composite's remote test site at the Mojave desert facility in California. The fiery accident occurred during the testing of components for the hybrid rocket engine that would be used on the next-generation rocket plane the company is building for Virgin Galactic. It was reported that something went wrong with the tanks of nitrous oxide being used to test SpaceShipTwo's motor and the blast of pressurized gas "went off like a bomb." Three workers were killed in the explosion and three others were severely injured.

Mr. Rutan told a press conference that the cause of the explosion was a mystery. The explosion happened during tests of the flow of pressured nitrous oxide through an injector. "We were doing a test we believe was safe. We don't know why it exploded," he said. [8]

A full investigation took place and the tragic accident delayed the announced plans of Virgin Galactic to provide private space flights from 2009. At the time, the president of Virgin Galactic, Will Whitehorn, expressed the company's sympathies to the families but would not comment further until Scaled Composites had completed its investigation.

One of the reasons for Branson's confidence in SpaceShipTwo is because of Rutan's 'trademark' common sense and safety first approach. This was demonstrated in his cleverly designed White Knight and SpaceShipOne, which performed brilliantly to claim the Ansari X-Prize just before the prize deadline expired.

Paul Allen spent a great deal of money to claim the prize, but what he actually bought was a place in aviation and spaceflight history. Apparently Allen considered it a bargain.
In an interview, Rutan said:

> "We have shown that you can construct a vehicle like this with a modest budget. The big question is, how many people will sign up, and will they pay $50,000 to $200,000 to go on one of those flights. It's not something I would contemplate unless I had partners willing to share the risk ... right now we are doing all these things as experimental flights. We have permission from the FAA to go Mach 2.5 straight up. There aren't many vehicles that do that. I think it will be good for the government to encourage something like space tourism. Having a space-tourism experience, whether sub-orbital or orbital, within the reach of people would be an exciting prospect."[9]

Let's put Rutan and Allen's accomplishment in perspective. From the 1970s through the early 2000s NASA spent many hundreds of millions if not billions on space plane development with excruciatingly poor results. NASA managed to cancel contracts for the development of a half dozen space planes or escape vehicles over three decades of failure. If you want to keep count, these were the HL-20, X-33, X-34, X-35, X-37, X-38 and X-43 – actually a baker's half dozen of seven.

Figure 2.6: The X-43, the latest in a series of space plane developments that NASA has canceled
(Graphic courtesy of NASA)

When the Rutan / Allen team claimed the X-prize, NASA emerged with a red face to match the red ink of its own attempts. One impromptu sign at the landing site at the Edwards Air Force base in California summed up the embarrassment with this succinct assessment: "SpaceShipOne – NASA Zero." This is why many feel that the time has arrived to allow entrepreneurs to come to the fore and let the space agencies play a supporting role when it comes to low earth orbit and space plane development.

Although this experimental sub-orbital space flight system performed well enough to claim the prize, it was certainly not a problem-free program and the stability of the SpaceShipOne vehicle was sufficient to delay several of the test flights. The greatest challenge now is to convert the experimental test vehicle into a craft that is reliable enough to fly "space tourists" into space on a routine and chartered basis. In order to move to this next plateau, a new company, the SpaceShip Corporation, was formed by Rutan and Allen with the charter to design and build SpaceShipTwo. And as described earlier, this is the corporation also backed by Sir Richard Branson, the founder of Virgin Galactic, and which is building his fleet of SpaceShipTwo vehicles.

Big Business Moves In

There was a major development in July 2007 when the Los Angeles aerospace giant Northrop Grumman Corp. agreed to buy Burt Rutan's company, Scaled Composites LLC. According to the Los Angeles Business Journal, the deal pairs an aerospace powerhouse with an emerging player in the nascent space tourism industry whose highly respected technology is advancing at a steady clip. But the acquisition has some wondering if a $30-billion company like Northrop, whose primary business includes shipbuilding and aircraft technology, will be interested in continuing with Scaled Composites' ambitious agenda of advancing the science of private manned rockets.

Northrop reported $3.3-billion of revenue in 2006 from its space technology division. About 90 percent of that came from government contracts but after the company lost a bid to build the successor to NASA's space shuttle, the company has not announced plans to build manned rockets. The company is, however, developing unmanned vehicles, as well as satellite and space radar technologies. Northrop spokesman Dan McClain declined to characterize Northrop's interest in space tourism. However, he said the company was not seeking any specific technology by buying Scaled Composites. He also said the deal would not affect Scaled Composites' operations and its entire management team would remain intact.

"We really value their current mode of operation," he said. "Northrop has always valued the innovative and entrepreneurial qualities of Scaled Composites and we think it's a good fit with our company's ongoing efforts with aeronautics and space flight."

Northrop first invested in the Rutan company in 2000, but through this transaction has agreed to increase its stake from 40 percent to 100 percent. Burt Rutan said:

> "My company has been owned by other corporations in the past and we have maintained our research and development culture that does the most efficient work in prototypes, and we expect no changes under the most recent equity revisions."[10]

Allen and the Birth of Microsoft[11]

As the financial strength behind Burt Rutan's enterprise, Paul Allen has played a critical role in moving the space tourism business from science fiction to science fact. But the story of his success and wealth began nearly 40 years ago. It was in 1968, at Lakeside School, a prestigious private school in Seattle, that Allen met eighth-grader Bill Gates who, like Allen, spent most of his free time figuring out the inner workings of their school's new computer. "Our friendship started after the mothers' club paid to put a computer terminal in the school in 1968," Gates told Fortune in 1995. "The notion was that, of course, the teachers would figure out this computer thing and then teach it to the students. But that didn't happen. It was the other way around." The pair became so adept with computer technology that, while still in school, they were both invited to serve as amateur technicians at a local computer center in exchange for free computer time.

Allen graduated from high school in 1971 and entered Washington State University. That same year, he read about the Intel Corporation's 4004 chip, the first computer microprocessor. In 1972, he and Gates purchased the next generation of the chip, the 8008, for $360. The pair used the chip to develop a special computer that conducted traffic-volume-count analysis and started a company called Traf-O-Data, planning to sell the computers to traffic departments. They eventually abandoned the company, and in 1974, Gates left Washington for Harvard University in Cambridge, Massachusetts. Allen followed, dropping out of Washington State and accepting a job as a computer programmer at the Honeywell Corporation in Boston. Allen hit upon the seed for their next business move in a *Popular Electronics* magazine cover story describing the MIT's Altair 8800 minicomputer. Recognizing that the computer would need a programming language, Allen and Gates set out to write a version of BASIC, a widely used computer language, specifically geared toward the Altair. They convinced MIT to buy their programming language. According to the *Fortune* interview, the credit line in the source code of their first product read: "Micro-Soft BASIC. Bill Gates wrote a lot of stuff; Paul Allen wrote the rest."

Allen and Gates soon changed their company's name to Microsoft and moved their business to Albuquerque, New Mexico. The young entrepreneurs quickly built up an impressive client list that included Ricoh, Texas Instruments, Radio Shack and another new start-up, Apple Computers. In 1978, with sales already over $1-million, they relocated their company to Seattle where, by 1979, they hired more than 35 employees and a professional manager.

The company entered into one of the most significant business deals in its history when International Business Machines (IBM) approached Microsoft seeking a programming language for its new personal computer, secretly under development. That same year, Allen negotiated the purchase of Q-DOS, a little-used operating system produced by Seattle Computer. Microsoft paid $50,000 for Q-DOS and, in turn, licensed the product to IBM for use with its new PC. In addition, Gates and Allen convinced IBM to allow other companies to copy the specifications of their PC, spurring the ensuing flood of PC "clones." The widespread availability of PCs necessitated compatible software programs which, in turn, required a universal operating system. The rest is largely history.

Hit by Cancer

In 1982, Allen was diagnosed with Hodgkin's disease, a form of cancer. He continued to work part-time at Microsoft during 22 months of radiation treatments, but in March 1983 he retired from the company and spent the next two years traveling, scuba diving, yachting, skiing and spending time with his family. He retained a 13 percent share of the company and continued to serve on its board. Finances were not a concern – as of 2003, he was estimated to be the world's fourth wealthiest citizen worth $21-billion – and Allen sought out new business and investment opportunities. These included Asymetrix, which produced applications that allowed both programmers and non-programmers to develop their own software, and Vulcan Ventures, an investment firm focused on technology. Allen also invested in numerous companies, including Ticketmaster, America Online, Egghead Software and the pharmaceutical company Darwin Molecular Corporation. Later investments focused on cable television, wireless modems, and Web portals.

In 1992, Allen founded Interval Research, a think tank focused on the Internet and compatible technologies. He also began to channel funds into entertainment and sporting ventures. He purchased the National Basketball Association's Portland Trailblazers in 1988 and the National Football League's Seattle Seahawks in 1996. He also purchased a reported 24% share of the film and television studio Dreamworks SKG. And internationally, he is following in the footsteps of Russian, Asian and other American billionaires by negotiating for the purchase of a major U.K. soccer team.

Allen celebrated an even earlier passion with his support of the Science Fiction Museum and Hall of Fame, which opened in Seattle in 2004. He further indulged his interest in other-worldly phenomena with a $13.5-million donation to the Search for Extra-Terrestrial Intelligence. Then he funded the development of SpaceShipOne, the world's first private spaceship, with a reported $100m. And now, along with Burt Rutan and Richard Branson, he is a key investor in the Spaceship Corporation that is building the SpaceShipTwo vehicles.

"Amazon Man" creates "Blue Origin"

Another space industry billionaire from the Internet era is Jeffrey Bezos, the founder and CEO of Amazon.com. His ground-breaking dot.com company is consistently ranked as one of the top retail sites on the Internet and offers over one million book titles via its web site, and now a dizzying array of other consumer products. His new company, Blue Origin, was created in 2000 with headquarters in Kent, Washington and with launch and test facilities in Culbertson County, a remote part of North Texas, very near the New Mexico and Texas border. The objective of Blue Origin is to start offering sub-orbital space tourism flights beginning in 2010, but the slogan of the company suggests a more ambitious goal, namely: "Creating an enduring human presence in space."

The company is developing a launch system designed to take off and land vertically. The first test flight of a development vehicle for the company's New Shepard spacecraft was in November, 2006. Blue Origin disclosed few details of this event, as has been the case with this secrecy-shrouded project to date. But apparently it did not achieve the desired test flight results or the needed speeds for even a sub-orbital flight. It was reported that this very short test firing rose to the height of the Washington Monument (about 87-metres) – and then landed.[12]

Figure 2.7: The New Shepard development vehicle prepares for its test launch.
(Photo courtesy of Blue Origin)

According to documents Blue Origin submitted to the FAA during 2006, the New Shepard Reusable Launch Vehicle would be cone-shaped, and stand about 15.25 meters (or 50 feet tall) and about 7 meters (or 22 feet wide) at the base. The fuel system is thought to be kerosene and hydrogen peroxide. The vehicle, for safety and escape purposes, consists of two stacked modules. One is to provide system propulsion, and the other provides escape capability for the flight crew.

At his Blue Origin web site, Bezos comments:

"We're working, patiently and step-by-step, to lower the cost of spaceflight so that many people can afford to go ... and so that we humans can better continue exploring the solar system. Accomplishing this mission will take a long time, and we're working on it methodically. We believe in incremental improvement and in keeping investments at a pace that's sustainable."[13]

Bezos continues to be active with the computer and IT entrepreneurial investment communities. This includes a leading role that he plays with regard to Flight School 05 and the X-Prize Foundation community where he has spearheaded efforts to create a viable space tourism business and encourage space plane development. Bezos told *Reuters* in late 2006 that his company eventually hopes to progress to providing passengers with not only sub-orbital flights but actual access to orbit. The apparent design of the New Shepard spacecraft entails that operations will be controlled completely by on-board computers, without ground control. Unmanned test flights, that began in November 2006, are expected to continue during 2007. Once passenger flights begin, the company has indicated that it expects to operate up to 52 launches a year – or once a week flights. Apparently the New Shepard vehicle is being designed to carry three or more people. The spacecraft in terms of basic shape and propulsion systems is closely modeled on the McDonnell Douglas DC-X derivative DC-XA vehicle, also known as the Delta Clipper.[14] This design was first developed and tested in the early 1990s but then canceled when NASA decided to back a reusable space plane design from the Lockheed Martin corporation instead. (If you count the Delta Clipper, NASA's ill-fated space plane projects listed earlier total eight.)

Bezos Sees an Opportunity

Jeffrey P. Bezos was born in Albuquerque, New Mexico where he spent most summers working on the ranch. At an early age, he displayed a striking mechanical aptitude; even as a toddler, he showed a remarkable talent by dismantling his crib with a screwdriver.[15] The family moved to Miami, Florida and at high school, he fell in love with computers and went on to Princeton University where he graduated with a degree in computer science and electrical engineering.

In 1986, Bezos joined FITEL, a high-tech start-up company in New York. Then, in 1988, he joined Bankers Trust Company, New York, leading the development of computer systems that helped manage $250-plus billion in assets and becoming their youngest vice president in February 1990. From 1990 to 1994, he helped build one of the most technically sophisticated and successful quantitative hedge funds on Wall Street for D.E. Shaw & Co., New York, becoming a senior vice president in 1992. Then he found the Internet.

Bezos learned early in the game about the Internet. This computer network was created by the US Defense Department's Advanced Research Projects Agency (ARPA) and was first known, not too surprisingly, as ARPANET. The concept was to create a maze of interconnected networks that would stay connected during an emergency or surprise attack even if many of the connections should fail. In a surprisingly short period of time, it was adopted by government scientists and academics at research universities to exchange data and messages. In 1994, however, there was still no Internet commerce to speak of and Jeffrey Bezos observed that Internet usage was increasing by 2,300 percent a year and growing, not only in the US but also around the world. He saw an opportunity for a new sphere of commerce, and immediately began considering the possibilities. In typically methodical fashion, he reviewed the top 20 mail order businesses, and asked himself which could be conducted more efficiently over the Internet than by traditional means. He identified books as a commodity where no comprehensive mail order catalogue existed, because any such catalogue would be too big to mail. He thus concluded that books, as well as music, were perfect for the Internet since he could create a sales company that would be able to share a vast database with a virtually limitless number of people that would only expand as the Internet grew across the world.

Figure 2.8: Jeff Bezos in celebratory mood
(Photo courtesy of Blue Origin)

He flew to Los Angeles to attend the American Booksellers' Convention and found that the major book wholesalers had already compiled electronic lists of their inventories. All that was needed was a single location on the Internet, where the book-buying public could search the available stock and place orders directly. Bezos knew the only way to seize the opportunity was to go into business for himself. It would mean sacrificing a secure position in New York, but he and his wife, Mackenzie, decided to make the leap.

They set up shop in a two-bedroom house, with extension cords running to the garage and three Sun microstations. When the test site was up and running, he asked 300 friends and acquaintances to test it. The code worked seamlessly across different computer platforms. On July 16, 1995, Bezos opened his site to the world, and told his 300 beta testers to spread the word. In 30 days, with no press, Amazon had sold books in all 50 states and 45 foreign countries (he called the company Amazon after the seemingly endless South American river with its almost infinite number of branches.)

The business grew faster than Bezos or anyone else had ever imagined. The company went public in 1997 and two years later, the market value of shares in Amazon was greater than that of its two biggest retail competitors combined. Amazon moved into music CDs, videos, toys, electronics and more. When the Internet's stock market bubble burst, Amazon re-structured, and while other dot.com start-ups evaporated, Amazon even began to post profits.

Today, Jeff Bezos and Mackenzie live north of Seattle and are increasingly concerned with philanthropic activities. Bezos has consistently been a person of vision and his new dream of moving people into space is fully consistent with his entrepreneurial view of the world. He consistently portrays the type of person who looks at the world and instead of asking why, asks "why not?"

Hotels in Space?

The process of creating private space ventures to achieve low cost and safe access to space is being pursued by a surprising number of space business enthusiasts. Another of them is the Budget Suites hotel magnate-turned-space-explorer Robert Bigelow. His effort to deploy an inflatable "spacehab" has already gone from science fiction to clear credibility. His first trial launch of a miniature prototype system occurred successfully in July 2006. The ultimate goal is to deploy an inflatable spacehab, three storys tall, in low earth orbit that will have greater volume than the International Space Station (ISS). These inflatable bio-systems will represent a small fraction of the cost and mass of the dense, metal-based ISS platform.

Robert T. Bigelow is a native of Las Vegas, Nevada and he graduated from Arizona State University with a Bachelor's degree in Business Administration. For over thirty years, the scope of his business endeavors included banking operations, real estate acquisitions, real estate brokerage and the management of extensive real estate inventory in the southwest US He is also the owner and developer of Budget Suites, a nationwide hotel chain.

Figure 2.9: Robert Bigelow – Planning hotels in space
(Photo courtesy of Bigelow Aerospace)

In 1999, he founded Bigelow Aerospace, described as a general contracting, investment, research and development company that seeks to achieve economic breakthroughs in space. His objective is to design, develop, fabricate and deploy habitable commercial space complexes at viable costs. He has been personally granted eight patents and has 13 more that have been submitted and are awaiting approval. He has already invested many tens of millions toward R&D and says he is prepared to invest a further $500-million by 2015 to realize his goals.

Bigelow Aerospace launched a 33%-scale model prototype Genesis I on July 12, 2006. It was 2-metres in diameter and the final version with thus be some 27 times larger in volume (i.e. 3x3x3). The second flight of Genesis II took place in June 2007 and the deployment of the first habitable commercial space structure, tentatively called the CSS (Commercial Space Station) *Skywalker*, is targeted for 2010. Like its predecessor a year earlier, Genesis II flew into space on a Dnepr rocket, a converted ballistic missile from Russia's military arsenal, and went into a near-circular orbit about 350 miles high. Compressed air from several on-board gas tanks began inflating the module shortly after arriving in space. Genesis II is more than 14 feet long and eight feet in diameter when fully expanded, according to Bigelow Aerospace. It carries an upgraded suite of internal avionics, a revamped inflation system and the various objects contributed by paying customers. Twenty-two cameras mounted both inside and outside the module are beaming back imagery to Bigelow ground stations in Las Vegas, Virginia, Alaska and Hawaii. Genesis II also carries Biobox, an animal habitat housing colonies of ants, cockroaches and scorpions.

The launch of Genesis was the culmination of a long dream. The whole time Bigelow was building hotels in Arizona, Nevada and Texas, he was secretly hoping to build in outer space. Then, seven years ago, he told his wife that he was starting a new company, Bigelow Aerospace. She thought: "Well, okay, we'll see. You know, it might be a passing fancy."

It wasn't. Bigelow now has a giant facility in Las Vegas and more than 100 employees. His goal is to build an orbiting complex of rooms that can be used by private companies, foreign countries or tourists – a type of public space station. He licensed the inflatable technology from NASA. "We stand on the shoulders of a NASA program that was canceled called TransHab," he says. TransHab was originally proposed in 1997 as possible crew quarters for NASA's space station. Congress eventually told NASA to stick with traditional aluminum cylinders, rather than spending money to develop inflatable technology. So once again we see with Bigelow Aerospace the willingness of entrepreneurs to pursue new, innovative and more cost effective technology that NASA or other traditional space agencies have abandoned.

Bigelow says inflatables have advantages. He says his balloons are actually less likely to get damaged by space debris, because the shell is made out of layers of tough materials, like a bulletproof vest. And it's much cheaper to launch something small that expands, rather than launching a big, rigid structure.

Over the next five years, Bigelow plans to launch several scale-modeled version of these inflatable spacecraft. And then he expects to be ready to launch the full-scale, three storys-tall capsule. George Whitesides of the National Space Society, a group that promotes space exploration, says the full-scale spacehabs are impressive. "I've been inside the full-scale versions of Bigelow's space hotel, and it is huge," Whitesides said. "Other entrepreneurs are building small rockets and spaceships that will give people a way of getting into space. But only Bigelow is working on building somewhere that people can stay and enjoy it."[16]

This, however, is not completely true. Although he is the current leader in the field, Bigelow is not the only pioneer pursuing the development of spacehabs. Inter Orbital Systems Inc. (IOS) is developing a large scale, one-and-a-half stage to orbit vehicle that would use the evacuated oxygen tanks as a habitat for space tourists once the rocket is in low earth orbit.

In addition to space tourism and orbital hotels, other expected uses for Bigelow's expandable modules include microgravity research and development and space manufacturing. The company plans to sell its BA 330 modules for $100-million apiece. He told Space News Business Report that he has invested more than $90- million in Bigelow Aerospace. "As a general contractor for 35 years," he said, "we're not strangers to contracting, to banking, to the financing of major projects. That's crucial if you really want to get the financial horsepower involved. Number one, the business model has to serve a customer. Number two, it has to be very cost-effective and number three, it has got to do what it says it's going to do. The banking world appreciates that and they respond ... Wall Street responds in predictable ways."[17]

Bigelow also that said the company's business structure will not only support destinations in low Earth orbit, but also operations on the Moon and to Mars.

The Next Big Prize – $50-million
Robert Bigelow has also announced a bold move into the space competition business with the "America's Space Prize," which is aiming to supports the development of new technology for going into space for the longer term. The award amount is set at the very compelling level of $50-million.[18]

Not surprisingly, the challenge to become the next Burt Rutan and claim a multi-million dollar prize, is much greater than the Ansari X-Prize. The winner must be a US-based entity, and develop the winning spacecraft without government funding – although use of governmental test facilities is allowed. The extremely difficult challenge involves building and flying a manned spacecraft that is capable of reaching an orbit of at least 400-kilometers altitude plus demonstrating the capability of carrying a crew of five and docking with a space station. The craft must then return safely to Earth and repeat the accomplishment again within 60 days. The spacecraft must complete at least two orbits of earth at the 400-kilometer altitude on both missions. The first flight must show that a crew of five can be carried, but the second one must actually carry a crew of five. Further, the spacecraft must essentially be reusable and very precisely no more than 20% of the craft can be expendable.

Perhaps most daunting of all, the deadline for claiming the prize is January 10, 2010. It is clear from analyzing the rules, or simply by talking to Robert Bigelow, that he is asking for a commercial space venture to develop an alternative capability to the Russian Soyuz craft and doing it within the next three years. As the concept of the prize competition developed, Robert Bigelow tried to get NASA to co-sponsor this challenge to the emerging US private commercial space industry. But due to regulatory constraints and other reasons NASA did not participate.

Bigelow, ever the entrepreneur who invests and risks his money cleverly, is apparently purchasing (for a million dollars or so) insurance coverage to protect against the $50-million payout. At this stage, the odds of successfully completing all of the requirements within the performance period seem small. Anyone seeking to become an official enrollee in the competition is required to register with Bigelow Aerospace and receive the detailed itemized rules.

In addition to offering the $50-million prize, Bigelow said his company is also is prepared to offer $200-million in conditional purchase agreements for six flights of the selected vehicle. He explained this concept as a part of his ultimate plan to deploy a private space station for space tourism and sees such a reliable private vehicle as critical to his mission. Bigelow has explained: "It could be somebody who doesn't win the competition, but who comes in late, but we like their architecture better than the winner's architecture."[19]

Bigelow Aerospace has received several honors for its spaceflight efforts. On October 3, 2006, Bigelow received the "Innovator Award" from the Arthur C. Clarke Foundation.[20] The award recognizes "initiatives or new inventions that have had recent impact on or hold particular promise for satellite communications and society, and stand as distinguished examples of innovative thinking." Robert Bigelow was presented with the award at the Arthur C. Clarke Awards in Washington D.C. alongside Walter Cronkite, who was honored on the same night with the Arthur C. Clarke Lifetime Achievement Award – an American icon appearing together with an icon to be.

"A Truck – not a Ferrari"

Elon Musk, another Internet entrepreneur, has taken a different approach to developing a new space capability with his SpaceX company. Born in South Africa, he taught himself how to write computer code and aged 12, he created a game he called "Blast Star" – which he sold to a computer magazine for $500. He felt isolated from the burgeoning software industry and when he was 17, he moved first to Canada and then entered the University of Pennsylvania to earn degrees in physics and business. In 1995, he was ready to begin his doctorate at Stanford when he was attracted by the Internet furore going on around him in Silicon Valley. He looked for a business that would generate cash quickly and founded Zip2, which helps newspaper companies put classified ads and other local information on web sites. After selling Zip2 to for $307-million, he started his next business, X.com, which offered a variety of banking services to consumers, but the most popular feature was the ability to e-mail money. Later, X.com acquired the rights to the name "PayPal" and focused on improving its e-mail money feature. PayPal soon became a favorite among users of eBay to make payments and just months after PayPal went public, eBay bought it, for $1.5- billion.

Musk was suddenly a billionaire in search of his next challenge – and space became his newest frontier.

Figure 2.10: Elon Musk – winner of NASA contract for SpaceX
(Photo courtesy of NASA spaceflight.com)

He had no cosmic experience but he started SpaceX, or Space Exploration Technologies, to build a rocket that he says will cost only a third as much as current models. He wants to see governments, universities and businesses that need an inexpensive way to get satellites into space turn to SpaceX. Musk is completely funding SpaceX himself and places his hopes on a 68-foot tall rocket called the Falcon Explorer. He says the Falcon will cost just $6-million to launch, which is a third of the going rate, because, he says, the Falcon is designed to be "a truck, not a Ferrari." This very low cost has attracted the US Air Force as a backer. To get the job done, Musk is luring top talent from large defense companies who have experience of building rockets. Tom Mueller, the company's vice president of propulsion, used to head liquid rocket propulsion development at TRW Space & Electronics.

Although his plans may sometimes seem crazy, he has already had some limited successes. The first stage of the two-stage rocket was successfully fired at a 300-acre testing facility in McGregor, Texas, and then in March 2006 he launched a Falcon that made its way almost to orbit before it experienced a failure. A second test flight in March 2007 also failed to reach orbit, but Musk apparently put a good spin on the failure, saying that the problem would be "pretty straightforward to address – we feel like there's really no need for an extra test flight."[21]

The third SpaceX launch in August 2008 was the first to carry a payload – three small satellites. One, called Trailblazer, was for the Department of Defense, and the two others were for NASA. These other two payloads were PRESat, a small automated laboratory, and NanoSail-D, intended to test of the concept of using sunlight to push a thin solar sail. This time, the launch failed following the second-stage ignition of the Falcon 1 rocket. The satellites were lost, together with a capsule containing the ashes of 208 people who had paid to have their remains shot into space. The ashes aboard the craft, included those of astronaut Gordon Cooper and the actor James Doohan, who played Montgomery "Scotty" Scott, the wily engineer on the original "Star Trek" television series. This unique and less than successful service was arranged by a company called Celestis, Inc. The ashes were widely distributed but certainly not as expected.

In spite of this third setback, Musk pressed ahead and in September 2008, Falcon 1 made a successful flight and became the first privately-developed rocket to achieve earth orbit. This time, SpaceX took no chances with a customer's payload and instead launched what it called a payload mass simulator – a 364-pound weight – from the Kwajalein Atoll in the central Pacific Ocean.

In a news conference after the launching, Musk told reporters, "It's great to have this giant monkey off my back."

Musk also stated that the fifth Falcon flight is being prepared and that he has given the go-ahead to begin fabrication of the vehicle for flight 6.[22]

The SpaceX company was the co-winner in 2007, along with Rocketplane Kistler (RpK), of the NASA COTS competition (i.e. Commercial Orbital Transportation Services). And following the successful Falcon 1 launch, Michael Griffin, the administrator of NASA, said "I am tremendously pleased for them. Practical commercial spaceflight remains a difficult goal, but one brought much closer with this step."

The COTS award means that Space X is contractually committed to provide a human-rated shuttle capability to and from the International Space Station (ISS) in the post-2010 time frame, after the Space Shuttle is permanently grounded. This R&D contract is worth nearly a quarter of a billion dollars.

As described earlier, RpK failed to meet the NASA-set targets in late 2007, but Space X remains under the gun to develop a reliable "man-rated" vehicle. It is now competing with Orbital Sciences Corporation (OSC), which was named by NASA in February 2008 to replace RpK in the COTS program.

Musk's initial backing and early launch commitments came from the US Air Force, which saw his Falcon-class launch vehicles as a means of addressing what the Department of Defense has labeled as the Operational Responsive Space (ORS) market. Although their rocket motors have performed well on static ground tests, the initial Falcon vehicles have yet to be proven on actual launches. However, Musk expressed his optimism in an interview with The Space Review. "There's likely to be a market there for personal spaceflight," he said. "I think over time the human spaceflight market will be much larger than the satellite market. It's just hard to say how long it will take for that to develop."[23]

The Ansari Family from Dubai

Anousheh Ansari was not only the first female fare-paying space tourist in 2006. She is also a member of a family determined to play a leading role in the privatisation of space flight. The Dubai-based family business supported the founders of the X-Prize competition to fund the award. Thus the competition became the Ansari X-Prize. As entrepreneurs, however, the Ansaris asked why put millions at risk? Let's buy insurance against what is seen as a long shot achievement, and it was the insurance companies that paid out the $10-million.

The Ansari's Plano, Texas, based company, Prodea, has invested in space-related activities including the establishment of a spaceport in the United Arab Emirates. Prodea is built upon the wealth of Iranian-American entrepreneurs Hamid, Anousheh and Amir Ansari, who made their fortune in the telecommunications industry. Prodea is also part of a 2006 business arrangement involving Space Adventures Inc. and the Russian space agency, Roskosmos, for the development of a fleet of "Explorer" spacecraft. The Explorer is based on the C-21 concept produced by Russia's Myasishchev Design Bureau for Space Adventures. This rocket plane would be flown up by a carrier aircraft, then launched to the edge of space – in much the same way as SpaceShipOne.[24]

Earlier in her career, Anousheh Ansari held positions with MCI Telecommunications Corporation and with Communication Satellite Corporation (COMSAT) in various engineering capacities. She then went on to create Telecom Technologies, Inc. (TTI), in Texas, where she is president and CEO. She was recognized by Working Woman magazine as the winner of the 2000 National Entrepreneurial Excellence award, and was

Figure 2.11: Anousheh Ansari training for weightlessness in Star City, Russia.
(Photo copyright 2006 Prodea Systems Inc.)

chosen as the winner of the 1999 Ernst and Young Entrepreneur of the Year, Southwest Region, for the Technology and Communications category. She has authored numerous technical papers and has two patents for her work on Automated Operator Services and Wireless Service Node. She holds a Master of Science Degree in Electrical Engineering from George Washington University and a Bachelor of Science Degree in Electrical Engineering and Computer Science from George Mason University.

She says she dreamed of being an astronaut while growing up in her native Tehran, Iran – a dream which came true for her in 2006, as described in Chapter 1.

The Key Role of Space Adventures
The company that helped put the first millionaires into orbit, in partnership with the Russia's Federal Space Agency (Roskosmos), is the Virginia-based Space Adventures. Now they are working with Russia's space agency, the Myasishchev Design Bureau and the Ansari company, Prodea, on the development of a fleet of Explorer spacecraft as described earlier. At altitudes in excess of 62 miles (100 kilometers), up to five people would experience a few minutes of weightlessness aboard the Explorer and see the curving Earth beneath the blackness of space.

Space Adventures' president and chief executive officer, Eric Anderson, has said spaceports to operate the Explorer could be located not only in the United States but around the world. Possible locations that have been identified include Singapore, Australia, the United Arab Emirates, and Malaysia. Meanwhile, Space Adventures will continue to pursue marketing deals with other sub-orbital companies.

Anderson was born in Denver, Colorado, in 1974, and he has helped to "invent" the space tourism business. The former aerospace engineer has sold over $100-million in thrill-ride tickets since founding the company in 1998. His company has flown more than 2,000 customers on a dozen different aircraft without incident. For $19,000 you get to see the blackness of space and the curvature of the Earth from the cockpit of a MiG-25 Foxbat hurtling at 2.5 times the speed of sound. For $7,000 you get ten 30-second bursts of weightlessness on the Ilyushin 76, a cargo plane Russian cosmonauts use for zero-gravity training. Zero-G, a Space Adventures affiliate, headed by X-Prize Founder Peter Diamandis, offers a weightless experience for just under $4,000.

Space Adventures has 20 full-time employees in its Virginia offices, and its founder Anderson owns a large part of it with several private investors owning the rest. It claims to have netted $1.5-million on revenues of $15-million last year – much of it from corporations like Oracle, American Express, Volkswagen and Citibank, which offer rides as customer and employee promotions. A NASA / Space Transport Association study predicts that the private space tourism market will reach $10-billion to $20-billion annually within a few decades. The key phrase here is "within a few decades."

More than a hundred clients are reported to have paid deposits for flights with Space Adventures and by late 2007, 26 of them had paid in full for sub-orbital flights, which guarantees

Figure 2.12: Space Adventures CEO Eric Anderson
(Photo courtesy of Space Adventures)

them a spot near the front of the line. (Their money is kept in escrow until they launch.)

The first "citizen astronauts" paid a reported $20-25 million to fly via the Space Adventures arrangements with Roskosmos. But now the price has increased. In a July, 2007, interview with *Space News*, Anderson said: "Actually, it's $30-million now. For the next couple of seats, that's the price." Among several other factors, the cost hike is due to the falling dollar – the rouble has appreciated some 50 percent against the dollar.

However, a possible change was signaled in April 2008 when Russia's space agency chief said the country may stop selling seats on its spacecraft to "tourists" starting in 2010 because of the planned expansion of the ISS crew from the current three to six or even nine. Anatoly Perminov said the space station's expansion will mean that Russia will have fewer extra seats available for tourists on its Soyuz spacecraft.

"We will continue flying tourists to the international space station in accordance with the existing programs, but we may have problems with it starting from 2010 because of planned increase of the ISS' crew," Perminov said, according to Russian news reports.[25]

Russia's space industries experienced hard times after the 1991 Soviet collapse when once-generous state funding dried up. They have survived mostly thanks to launches of foreign commercial satellites and revenue from so-called "space tourists." But more recently, Russia's oil-driven economic boom has led to increases in government spending on the nation's space program, reducing the space agency's dependence on revenue generated by commercial flights.

And Next – Spacewalks and Lunar Missions

Anderson has also announced that a spacewalk is a possibility on one of the future flights. The Space Adventures fee for a spacewalk is $15-million on top of the $30-million base price and Anderson added: "One of the consequences of the spacewalk is that you get a little bit more time up there, instead of a week to 10 days, you'll probably get close to three weeks." The private spacewalker would exit the ISS out of a Russian airlock, outfitted in an Orlan space suit.[26]

However, the upcoming private space flights must wait until after two ISS-bound Soyuz trips carrying astronauts for the countries of Malaysia and South Korea. Russia launched Malaysia's first spaceflyer to the ISS with the station's Expedition 16 crew in October 2007. South Korea's first astronaut will accompany two professional spaceflyers to the ISS during a planned spring 2008 launch. Seat availability for future fare-paying citizen astronauts may also be increased once the ISS grows from its current three-person capacity to a full six-person crew in 2009. "Access to the space station is a pretty valuable thing and it's a challenge," Anderson said. "All the clients who want to go, we'll find them seats. It may take some time but we'll get them up there."

So what next? A Space Adventures team has blueprinted a circumlunar mission using a unique blend of existing and flight-tested Russian technology. At the heart of the lunar leap is Russia's Soyuz spacecraft. A pilot and two passengers would depart Earth in their Soyuz, linking up in orbit with an un-piloted kick stage for a boost outward to the Moon. The project is called Deep Space Exploration (DSE)-Alpha and will be the first in a series of deep space missions being planned by Space Adventures at the cost of $100-million per person. The mission as now planned would be conducted in cooperation with the Russian Federal Space Agency (FSA) and the Russian space design bureau, Energia.

Eric Anderson said: "The Soyuz was originally designed as a circumlunar spacecraft. It hasn't flown with people around the Moon, of course. But the Soyuz would fly a free-return trajectory – a boomerang course – around the Moon. So there's not a lot that needs to be done to the Soyuz to accommodate for that ... it could probably fly around the Moon right now."[27]

Anderson's Early Days

Anderson began thinking hard about space tourism while interning at NASA in the summer of 1995. He was a junior in the aerospace engineering program at the University of Virginia and these internships taught him two tough lessons: one, his 20/40 eyesight would always keep him from taking the test to gain astronaut status, and two, you'll never be able to trust NASA-crats to figure out how to put civilians into space for a reasonable cost. "Things there are massively over-inflated," says Anderson. "I'd see a million-dollar study produce a 100-page report I could have written in college. The government has no incentive to make things cheaper. The bigger the budget, the more power they wield."

After graduating at the top of his class from the University of Virginia's engineering school in 1996, Anderson raised $250,000 from investors – including Peter Diamandis, founder of the X Prize Foundation, and Michael McDowell, who started the Arctic cruise company Quark Expeditions – to launched Space Adventures from a room in his Arlington townhouse. Inspired by McDowell's idea of selling adventure vacations aboard Russian icebreakers and submersibles Anderson figured he could connect rich thrill-seekers with a Russian government barely able to afford its space program. McDowell introduced him to military types in Moscow, but the skies were new territory, and the Russian bureaucracy proved a nightmare. "We found people who said they could arrange the flights," says Anderson, "only to learn later they had no authority. Half the battle was getting to the right person."

Once he had secured permission, he went about finding clients for flights in a Russian MiG fighter plane. At an Explorers Club gala in New York he got his first three: Lotsie Holton, an heiress to the Anheuser-Busch fortune, and her son and husband. The trio had such a good time on their MiG flights that they spread the word among the wealthy. Soon Anderson had full flights.

He had no idea he'd be able to sell tickets to the International Space Station until a February 2000 meeting with Dennis Tito at the offices of Tito's company, Wilshire Associates, in Santa Monica, Calif. The previous year Anderson had paid the Russians $100,000 to study whether the Soyuz could be used to transport tourists to the ISS. The Russians said it could be done, and Anderson happened to have the study's results on him when he stopped in to ask Tito to invest in Space Adventures.

"We talked a long time," recalls Anderson. "Then Tito said, 'I think what you're doing is great – it's the future – but I'm completely uninterested in investing in your company. But I do want to orbit the Earth now. Can you help?'"

So began ten months of intensive negotiations with the Russians, culminating in seven signatures on the final agreement. Anderson was at one of Tito's first meetings with the Russians, a typical 11 a.m. brunch at Star City, with bottle after bottle of vodka being consumed. After three hours the general in charge, now tipsy, suddenly announced it was time for Tito's medical exam. They went out to the vestibular chair, basically a barbershop seat that spins at 40 revolutions per minute for 15 minutes, while an engineer barks out commands like, "Move your hand forward, touch your chest," at the poor soul being spun. After Tito got off the chair without losing his brunch, the general said: "Mr. Tito, he's a real man. No problem for space flight."

Once Tito had flown and then successfully touched down in the Kazakhstan desert, Anderson felt euphoria and intense relief, but also a little envy. A big part of him wished he had been in Tito's shoes. "I'll go on a sub-orbital test flight in two or three years," he said. "I would never sell anything I wouldn't do myself."

Figure 2.13: John Carmack experiences ZeroG
(Photo courtesy of Armadillo Aerospace)

Video Game Leader and Rocketry Enthusiast

John D. Carmack II is a widely-recognized figure in the video game industry. Though he is best known for his innovations in 3D graphics, Carmack is also a rocketry enthusiast and the founder and lead engineer of Armadillo Aerospace. He has aspirations of suborbital space tourism in the short term, eventually leading to orbital space flights and he developed the Black Armadillo project to compete for the Ansari X-Prize.

Born in August, 1970, he became a prolific programmer and co-founded id Software, a computer game development company, in 1991. He was the lead programmer of the highly successful id computer games Wolfenstein 3D, Doom, Quake, and subsequent sequels to Doom and Quake. His revolutionary programming techniques, combined with the unique game designs of John Romero, led to a mass-popularization of the first-person shooter genre (FPS) in the 1990s.

It was exciting to move to a new field I didn't know anything about," Carmack said. "I am drawn to the engineering. I enjoy solving problems and finding novel solutions to things, and I've been at the top of my field in software for so long. The challenges, while they evolve, they are not so novel anymore."

Armadillo Aerospace was founded in 2000 and is based in Mesquite, Texas and its initial goal was to build a manned suborbital Ansari X-Prize-class spacecraft, but it has now stated long-term ambitions of orbital spaceflight. The company places a strong emphasis on a rapid build and test cycle. Armadillo Aerospace has designed and built a number of different vehicles using a variety of propellants. Each design has several features in common. One is the use of modern computer technologies and electronics to simplify rocket control and reduce development costs. Another is the use of liquid propellants and VTOL (Vertical Take-Off and Landing) to facilitate short launch-to-launch times.

Armadillo's X-Prize vehicle was unorthodox among modern rockets in that instead of using stabilization fins, which complicate the design and increase drag, it used an aerodynamically unstable design, where the computer-controlled jet vanes are based on feedback from fibre optic gyroscopes. A preference for simplicity and reliability over performance was also evident in its choice of hydrogen peroxide (50 % concentration in water) and methanol as a mixed monopropellant for the vehicle. Since the completion of the X-Prize, however, the Armadillo Aerospace designers have opted to switch to liquid oxygen because of difficulties with peroxide catalysts and the lack of availability of high-concentration peroxide in the United States for small companies.

In June 2004, Armadillo successfully demonstrated a computer-controlled VTOL flight of its prototype vehicle, becoming the third unmanned rocket in history to have done so, after the McDonnell Douglas DC-X and Japan's Institute of Space and Astronautical Science (ISAS) Reusable Vehicle Test (RVT).

Figure 2.14: The lunar lander's qualification flight at the Oklahoma Spaceport.
(Photo courtesy of Armadillo Aerospace)

Armadillo Aerospace competed in the 2006 X-Prize Cup where they were the only competitor in the NASA Lunar Lander prize challenge. The company took two similar vehicles, Pixel and Texel, to the event. The vehicles narrowly failed to win the Level 1 prize, after making three dramatic attempts totaling over 5 minutes in the air, but finally crashing out on the final attempt. Persistent landing problems were the main cause of failure, with the undercarriage breaking several times, and landing slightly off the pad on one occasion due to guidance issues.

However, in July 2007, Carmack reported successful progress in his group's modular rocket work by completing a set of tethered flights of the first modular rocket segment much quicker than expected. The next step is building four additional modules to be flown in one module, two module, and four module configurations, and then demonstrating two single module vehicles flying at the same time.[28]

First of the Computer Entrepreneurs

Jim Benson founded his SpaceDev company as long ago as 1997 and could be said to have started the trend of successful computer and Internet dot.com entrepreneurs moving into the space development arena. After a successful career in the computer industry, he never achieved billionaire status, but still amassed a fair amount of walking around money. He decided to take on the challenge of starting a space commercialization venture which combined his lifelong interests in science, technology and astronomy with his successful business experience.

Benson formed SpaceDev and worked to develop the world's first private sector enterprise to profitably explore and develop space beyond earth orbit. The company's mission was to help "make space happen" for all of humanity, through the development of a comprehensive private space program.

SpaceDev focused on developing its sub-orbital Dream Chaser™ which is derived from an existing X-Plane concept, as used in the SpaceShipOne. It will have an altitude goal of approximately 160 km (about 100 miles). It will be powered by a single, high performance hybrid rocket motor, under parallel development by SpaceDev for the SpaceDev Streaker™, a small, expendable launch vehicle designed to affordably deliver small satellites to low earth orbit. The Dream Chaser™ will use motor technology being developed for the Streaker™ booster stage, the most powerful motor in the Streaker family.

SpaceDev currently creates and sells a range of innovative space products, services and solutions to government and commercial enterprises. Its innovations include the design, manufacture, marketing and operation of micro- and nano- satellites, hybrid rocket-based orbital maneuvering and orbital transfer vehicles as well as sub-orbital and orbital hybrid rocket-based propulsion systems for human space flight. Under Benson's guidance, SpaceDev developed critical hybrid rocket motor technology and furnished all of the rocket motors for SpaceShipOne, the craft that earned the $10 million Ansari X Prize in 2004.

During his 10 years with SpaceDev, Benson served as founder, chairman, chief executive officer and chief technology officer. But in September, 2006, he stepped down to start a new venture called The Benson Space Company.

He has been an outspoken critic of NASA's efforts to design and build a space plane (i.e. the HL-20, the X-33 and follow-on projects.). He was quoted as saying shortly before leaving SpaceDev:

"I have been waiting for almost fifty years for commercial space flight … I have concluded that SpaceDev, through its unbroken string of successful space technology developments, now has the technical capability and know-how, along with our partners … to quickly develop a safe and affordable human space flight program, beginning with sub-orbital flights in the near future, and building up to reliable orbital public space transportation hopefully by the end of this decade."[29]

Benson is a founding member of the Personal Spaceflight Federation and was recently appointed to the Board of Directors of the California Space Authority. He founded the non-profit Space Development Institute, and introduced the Benson Prize for Amateur Discovery of Near Earth Objects. He is Vice-Chairman and private sector representative on NASA's National Space Grant Review Panel.

"I am dedicated to opening space for all of humanity," he said. "And, with SpaceDev well-managed and growing, I plan to spend the next several years creating the possibility that anyone who wants to go to space will be able to, safely and affordably." Sadly, James Benson died at his home in Poway, Ca in October 2008, aged only 63.

Figure 2.15: James Benson addressing the International Space Development Conference in Dallas in May 2007.
(Photo courtesy of J. Foust / The Space Review)

There Are More…

There are, of course, more than just billionaires and well-known entrepreneurs at work in this business. Many of the nearly 50 space plane developers and space tourism backers are operating out of garages and, for some, their capital financing plan involves "maxing" out as many credit cards as they can get. Some of them are described in the next chapter.

But, tomorrow's would-be space billionaires are intent on following in the footsteps of Paul Allen, Elon Musk and Richard Branson. They are diligently toiling away at developing their space planes designs and trying to make their dreams a reality.

Who knows, they may become the billionaires of tomorrow – or not? Undoubtedly, many will fail and indeed some have already done so. The current count of failed, folded or merged projects, as registered in Appendix B, totals some 25. In the next chapter, however, we will explore those whose dreams and ambitions may not reach to the stars but will certainly rise to the stratosphere.

References:
1. Time.com, 22 February 2007 – http://www.time.com/time/magazine/article/0,9171,1592834,00.html
2. Cathy Booth Thomas, "The Space Cowboys," Time, March 5, 2007, p58.
3. London Daily Telegraph, August 2008.
4. Will Whitehorn, London Independent, July 2008.
5. Richard Branson, Wikipedia (http://en.wikipedia.org/wiki/Richard_Branson) etc, etc.
6. RealityTV – http://realitytv.about.com/od/therebelbillionaire/a/BransonBio.htm
7. Virgin Group web site – http://www.virgin.com/AboutVirgin/RichardBranson/WhosRichardBranson.aspx
8. Daily Telegraph, UK –http://www.telegraph.co.uk/news/main.jhtml?xml=/news/2007/07/27/wvirgin127.xml
9. Newsweek – http://www.msnbc.msn.com/id/5251227/site/newsweek/
10. Los Angeles Business Journal, July 23, 2007
11. Paul Allen, Wikipedia – http://en.wikipedia.org/wiki/Paul_Allen
12. Leonard David – Space.com – 4 January 2007.
13. Space.com – January 4 2007 – http://www.space.com/missionlaunches/070104_bezos_blueorigin_updt.html
14. Blue Origin, Wikipedia – http://en.wikipedia.org/wiki/Blue_Origin
15. Academy of Achievement – www.achievement.org/autodoc/page/bez0bio-1

16. NPR – http://www.npr.org/templates/story/story.php?storyId=5555718
17. SpaceNews, March 26 2007 – http://www.space.com/spacenews/070326_bigelow_businessmonday.html
18. Leonard David, "Rules Set for $50 million "America's Space Prize" – Space News www.space.com/spacenews/businessmonday_bigelow_041108.html
19. Benson Starts Independent Space Company to Market SpaceDev's Dreamchaser, September 28, 2006 http://www.spacedev.com/newsite/templates/subpage_article.php?pid=583
20. Arthur C. Clarke Foundation – www.clarkefoundation.org
21. Space.com – http://www.space.com/missionlaunches/070320_spacex_falc1_test2.html
22. Schwartz, New York Times, 3 August 2008.
23. The Space Review, Nov. 14 2005 – http://www.thespacereview.com/article/497/1
24. MSNBC, February 2006 – http://www.msnbc.msn.com/id/11393569/
25. Vladamir Isachenkova – Associated Press – April 11, 2008
26. Space News, July 15 2007.
27. Space.com, August 10 1005 – http://www.space.com/news/050810_dse_alpha.html
28. Leonard David, Space.com – 27 July 2007
29. Benson Starts Independent Space Company to Market SpaceDev's Dreamchaser, September 28, 2006 http://www.spacedev.com/newsite/templates/subpage_article.php?pid=583

— Chapter 3 —

"Space entrepreneurs still tend to be seduced by the tendency to mistake technical possibility for market opportunity ... today's entrepreneurs represent a sophisticated and seasoned group of business managers who stand a better chance of navigating the many daunting technical and market challenges associated with the new commercial space industry sector ... the good news is that the US government is doing a better job of fostering a more stable and predictable regulatory and policy climate for space entrepreneurs." Courtney Stadd, former NASA & US Dept. of Transportation Executive.

THE LEADING PLAYERS AND THEIR CHALLENGERS

The billionaire entrepreneurs are not alone in pursuing space travel for the masses. There are nearly fifty companies around the world developing new spacecraft, planning futuristic spaceports or seeking to offer a range of "space travel" services. Many of these, however, will probably "drop out" later – some more likely sooner than later. Twenty-five such ventures, some of which were originally X-Prize hopefuls, have already faded from the scene in just the last three years.

If you want a good mental image of the space tourism "industry" today, here's a thought. Just think of the pioneering days of aviation at the start of the 20th century. Then play it fast forward. If you see a motley group of dare-devils, wealthy industrialists, inventors, and serious engineers all mixed together you are not far off the mark.

Some of the aspirants we have already met are billionaires with their own capital to burn and who have carefully chosen their teams of experts. Others are large-to-mid-size aerospace companies with tremendous technical capabilities at their command. Then there are would-be space tourism businesses that are true "start-ups," all mixed in with the seriously wealthy entrepreneurs. These most tenuous of the "rocketeer aspirants" are often operating on a shoestring from garages or university lab facilities. These are the space enthusiasts and dreamers seeking to make their mark on history. These avid pursuers of space commercialization very often have smart web sites, lofty business plans and great vision statements. They also have a very long shot at becoming 21st century success stories. Very, very few have the potential to become the latest version of a Hewlett-Packard or Microsoft by springing from obscurity to fame and fortune. Their goals are as stratospheric, but their odds of success are longer than a plow horse at the Kentucky Derby. These would-be rocketeers are long on hope and aspirations, but short on capital and world-class scientists and engineers.

The key to understanding the space tourism business is to note that most of the general public are not willing to risk their lives on "bleeding edge" technology. As Courtney Stadd, formerly of NASA and the Department of Space Commercialization, has noted: "Space entrepreneurs ... tend to be seduced by the tendency to mistake technical possibility for market opportunity..."[1]

The sophisticated upscale market, willing to pay $200,000 or more for a sub-orbital flight to the edge of space and back, wants security, glitz, and pampering – but most of all security and safety. The would-be spacecraft designers, operating out of makeshift labs and rickety garage facilities, maybe thinking like today's Wilbur and Orville Wright. But they are unlikely to provide either true breakthrough technology or reassuring safety and security features.

To date, more than a billion dollars has been invested in the new space tourism industry and another billion or more will be needed before this industry truly "takes off." We are years and gigadollars away from knowing if this incipient industry will truly succeed or fail. The odds will thus be on those with staying power and deep pockets rather than those with just a dream.

It is probable that less than a dozen or so of the ventures listed in Appendix A will ultimately deliver a viable product and offer a reasonably secure service that the high flying jet set market will embrace. Market validation – and consolidation – is the dominant name of the game (See Appendix B for a listing of those that have already faded from the scene. Probably some of the aspirants listed in Appendix A will soon follow suit.)

Among those described in the previous chapter, the present leader in the field appears to be The Spaceship Corporation, Inc. which was created in July 2005. This leading developer of sub-orbital flight launch system combines the capabilities and capital of Burt Rutan, head of Scaled Composites, Sir Richard Branson's Virgin Group, and Paul Allen. This amazing team is pursuing a plan to develop and build a new generation of sub-orbital spacecraft – and will continue to do so according to the aerospace giant Northrop-Grumman which acquired 100% of Rutan's company in mid-2007.

Richard Branson and Jeff Bezos have both ordered initial fleets of these craft, with the first test flights taking place in 2009. As described in the previous chapter, Bezos, the Amazon.com mogul, is developing his own rocket, the New Shepard. This vehicle is based on the Delta Clipper, but he has hedged his bets by also ordering his own fleet of Rutan's SpaceshipTwo vehicles to start his service.

Another major player, of course, is Elon Musk who is continuing with his flight tests of SpaceX designed Falcon vehicles. At this time, Space X – under separate R& D contracts – is being significantly backed by NASA's nearly half-billion dollar COTS program to create reliable commercial flight systems. This unconventional NASA research initiative aims to find a private operator to fly astronauts to and from the International Space Station in the post-2010 time frame when the Shuttle is to be retired. And clearly, if the vehicles developed by Space X and the other selected company, Orbital Sciences Corporation, can provide reliable commercial service to low earth orbit for NASA they can also operate space tourism flights as well.

Then there is computer gamer John Carmack's Armadillo Aerospace enterprise. Carmack has not only begun the development of the Black Armadillo launch system to support a space tourism business, but has also sought to develop a maneuverable vehicle to win NASA's Centennial challenge prize. This is a system with a number of vernier jets, well adapted to take off and land on the Moon and Mars with a high degree of precise control. The Black Armadillo still has a long way to go to prove its functionality as a space tourism vehicle.

Meanwhile Jim Benson, the founder of SpaceDev claimed that SpaceShipOne and SpaceShipTwo both depend on the rocket technology which his company developed. Benson had also been an outspoken critic of NASA's efforts and when we last talked to him in Hawaii at the Japan-US Science Technology and Space Applications Program, just before he stepped down from SpaceDev to his new spin-off Benson Space Corporation, he was still not pulling his punches. He accused NASA of disqualifying him for a major award because he bid prices too low. As Benson complained – what the heck is that all about? Benson claims that Burt Rutan and his crew would be nowhere without their innovative laughing gas and neoprene rubber fueled rocket motors. A cynic could respond, of course, that you might have a much cleaner craft than one that burns up rubber. Anyone that has burned old tires will know what we mean.

Then there is Eric Anderson of Space Adventures. He has placed orders for his own spacecraft and in doing so he has followed the advice of Yogi Berra: "When you come to a fork in the road, take it." In this case, Space Adventures is supporting a two-fold development. As option one, they are funding the Xerus spaceplane developed by the US based XCOR Corporation. And, in parallel, they are also working with the leading Russian aerospace design bureau (in tandem with the Russian Space Agency Roskosmos) to create a space plane that is a spin-off from an existing Russian design.

The following table (see Figure 3.1) is our assessment of the top contenders to deploy the first commercial space tourism flights. This is, of course, today's assessment, that could well change in coming months.

All of these companies are striving to get there first, and to stay the course. But success may not ultimately come to the wealthy entrepreneurs and the best-publicized projects. Many other companies and their leaders, as described in this chapter, believe they also have the experience, expertise and determination to play a key role in the development of space tourism. Hope springs eternal, and so forth...

Figure 3.2 (page 52) illustrates the wide range of technical approaches being taken by the various companies fervently seeking the most efficient and safest approach to conducting sub-orbital flights. Indeed some companies are seeking to develop craft capable of reaching low earth orbit and returning

Company Name	Vehicle	Technical Approach	Concerns	Start of Service
Armadillo Aerospace	Black Armadillo	1 stage. LOX/ethanol engine. (Limited capital investment). Vertical Takeoff and land. (Like the Delta Clipper design.)	New system. Limited tests	2010?
Blue Origin	Initially will depend on start-up Spaceship 2 fleet. Followed by New Shepherd launch system	Reusable Launch Vehicle. Hydrogen Peroxide and Kerosene fuel. Abort system.	New Shepherd is a developmental system. Limited tests	2010?
Rocketplane-Kistler	Rocketplane XP Pathfinder spaceplane plus K-1 rocket vehicle to LEO and to the ISS.	Sub-orbital. 4 seat fighter-sized vehicle. Up to 4 people or 410 kg to 100 Km. K-1 rocket -Payloads to LEO, MEO, GTO, ISS Cargo re-supply & return	New systems. Requiring test and verification for both spaceplanes.	Status very much up in the air after NASA contract canceled.
Space Adventures (with XCOR Corp)	XCOR vehicle	Xerus (Sub-orbital space) (HTHL) Isopropyl alcohol/LOX	New system. Limited tests	2009-2010?
Space Adventures (with Myasishchev Design Bureau)	Explorer Space Plane (C-21) lifted to high altitude by the MX-55 High Altitude launcher plane (HTHL)	Liquid fuel motors. Horizontal Takeoff and Horizontal Landing (HTHL) (lifting body with parachute landing)	Based on extension of Russian systems but still this is a new system	2009-2010?
SpaceDev (together with Benson Space Corporation)	Dreamchaser	Single Hybrid Engine. (Neoprene and NO_2) for sub-orbit. Launch of spaceplane on the side of 3 large hybrid boosters to reach LEO orbit & ISS.	Both sub-orbital and orbital system derive from SpaceShip 1 but still a new system.	2010?
Spaceship Corp.	Upgraded version of SpaceShipOne with increased cabin size. This is called SpaceShipTwo.	Hybrid Engine. (Neoprene and NO_2) for sub-orbit) Flown to high altitude on a jet based launcher system.	Expanded version of Space Ship 1, but still requiring extensive test. Likely to be first operational system for Virgin Galactic and others.	2009?
Space X	Dragon Space plane and Falcon 9 to low earth orbit	Both systems based on Falcon rocket technology. Liquid-fueled systems	Both spaceplane and rockets based on liquid-fueled Falcon system. Both require extensive tests.	2010?
Virgin Galactic	Fleet of Space Ship 2. Upgraded version of Spaceship 1 with increased cabin size	See. Spaceship Corp. above.	As first system. Will require earliest tests.	2009?

Figure 3.1: The Space Tourism Leaders

to Earth as well. Some of these innovators are seeking alternatives to the familiar blast-off used by NASA, the Russians, the Chinese, etc. where the crew and passengers in the spacecraft sit on top of a huge quantity of explosive fuel on the launch pad and make a fiery and thunderous departure to escape the Earth's gravity – not exactly a joy ride for Citizen Astronauts.

This diversity of approach and extended trial periods may end up with duplicative actions and wasted resources, but it also may also provide the clearest pathway to success. This open competitive laboratory will allow the best designs to rise to the top – literally and figuratively. We believe this open competitive process will ultimately prove invaluable to the development of new and proven passenger craft for the space tourism business. Testing dozens of approaches against one another in a global marketplace seems to be a better idea than prematurely locking on to a single unified design approach. NASA's efforts to develop space planes in the "conventional way" of choosing the design up front have been a disastrous failure.

Various Approaches for Accessing Space	Companies Using this Particular Approach
Lighter than Air Ascender Vehicles and Ion Engines with high altitude lift systems providing access to LEO.	JP Aerospace (Commercial venture with volunteer support)
Balloon Launched Rockets with capsule return to ocean by parachute	Da Vinci Project, Planetspace, HARC, IL Aerospace
Vertical Takeoff and Vertical Landing	Armadillo Aerospace, Blue Origin, DTI Associates, JAXA, Lockheed Martin/EADS, Masten Space
Vertical Takeoff and Horizontal Landing (spaceport)	Aera Space Tours, Bristol Space Planes, C&Space, Air Boss, Aerospace Inc., Energia, Lorrey Aerospace, Phoenix and Pre-X by EADS Space Transportation, Planetspace, SpaceDev, Space Transportation Corp., Space Exploration Technologies (SpaceX), Sub-Orbital Corp, Myasishchev Design Bureau, t/Space, TGV Rocket, Vela Technologies, Wickman Spacecraft & Propulsion.
Vertical Takeoff and Horizontal Landing (from ocean site)	Advent Launch Site, Rocketplane/Kistler (Financial Status unclear after NASA Contract cancelled)
Horizontal Takeoff and Horizontal Landing	Andrews, Scaled Composites, the Spaceship Corporation, Virgin Galactic, XCOR, and Project Enterprise by the TALIS Institute, DLR and the Swiss Propulsion Laboratory
Tow Launch and Horizontal Landing	Kelly Space & Technology Inc.
Vertical Launch to LEO from Spaceport	Alliant ATK, Inter Orbital Technologies, Rocketplane/Kistler, SpaceHab, UP Aerospace
Launch to Leo from Cargo Jet Drop	Triton Systems

Figure 3.2: Technical Design Approaches for US and International Private Commercial Space Systems

Over the past fifteen years, NASA has started and stopped a half dozen projects with marginal results. If this criticism of NASA seems unduly harsh, anyone can review the history of NASA's space plane developments, variously known as the HL-20, X-33, X-34, X-35, X-37, X-38 and X-43 projects. A company that embarked on seven different projects in a row and never brought one to market would no longer be in business.

In addition to the companies described earlier and in chapter 2, the space tourism and space plane businesses that are among the leaders at present are as follows. (Note that these are the US teams; we will cover the international market in the following chapter.)

Rocketplane-Kistler

This has been a very much up and down affair. Kistler Aerospace, after raising $600 million in 1998 went into chapter 11 in July 2003. In 2005 it re-emerged from bankruptcy and in February 2006 merged with the Oklahoma-based Rocketplane to form Rocketplane-Kistler (RpK) and many thought this would lead to smooth sailing. The new company, headquartered in Oklahoma City, was later chosen by NASA as one of the two operators of next-generation orbital and sub-orbital space transportation vehicles. Rocketplane, after the merger, was selected from 23 would-be providers in the NASA competitive process to be one of the two Commercial Orbital Transportation Service (COTS) providers. Winning this competition, and thus being in line to receive over $200-million in research funds from NASA to develop their systems, undoubtedly positioned RpK very strategically in this highly competitive field. However, as mentioned earlier, the company later failed to meet the financial milestones set by NASA to raise $500 million. Thus RpK was replaced when NASA canceled their COTS contract after doling out some $32-million. RpK appealed to the General Accountability Office to no avail and is now very much on the outside looking in.

The second company originally selected by NASA in the COTS competition was Elon Musk's SpaceX. Musk and his impressive start up company was described in Chapter 2.

Despite the NASA setback, RpK is still seeking to market the fully reusable, two-stage Kistler K-1 orbital launch vehicle that is to launch the fully reusable, sub-orbital Rocketplane XP Spaceplane. These vehicles are currently both slated to be fully operational in the next two years, but under-capitalization could prove to be a problem. Time will tell if they meet this schedule.[2]

It was an Oklahoma newspaper that said Rocketplane 'had seen more downs than ups' since its inception six years ago, but added that the company is still slated to eventually send people into space. Their project, which has received support from the state of Oklahoma as well as NASA, has not yet sent a reusable vehicle into space on flights more than 60 miles above Earth. In 2007, the company rolled out major engineering changes to its Rocketplane XP

Figure 3.3: Artist's impression of the Rocketplane XP spaceship
(Courtesy of Rocketplane Global)

suborbital vehicle. David Faulkner, the program manager, said the capacity of the craft has been increased to hold five rather than three passengers and with a move away from the earlier Learjet design to allow weight savings and more flexibility. Faulkner said the main focus is now on securing private financing to finish the project. The company did not disclose how much in private funds it needs to raise, but Faulkner said the project is still on track to one day offer regular people the chance to visit space, and said he is confident the needed funding will be secured. "Funding ebbs and flows just like any aerospace project," he said. "We've had to pull back a little so we can get to that next phase."[3]

Then, in July 2008, Rocketplane Global (like Virgin Galactic earlier) recognized the PR value of weddings in space with an event in Tokyo. The Space Wedding Charter Flight business was announced by First Advantage and Rocketplane Japan who said space wedding ceremonies would be available to customers all over the world who would like to have the ultimate "high-end" celebrity wedding. The special package includes the actual wedding ceremony in space for the bride, groom and three guests, the space marriage license and certificates, an original wedding dress, full picture and video coverage of the wedding flight and a live broadcast of the ceremony to the ground, premium hotels and transportation, an original web site developed for the wedding customers, and a premium concierge service hot line.

Rocketplane CEO George French said: "This innovative package shows how the XP spaceplane can be adapted for a wide variety of unique and special uses. We are developing a variety of additional charter flight packages involving artists, media and cultural themes as well as additional corporate promotional campaigns." He added that the "shirtsleeve cabin environment" of the XP spaceplane with its redundant life support systems is a major factor in this type of specialty charter flight business. The wedding party can therefore wear clothing of their own choosing, and not be encumbered with bulky helmets or pressure suits which would detract from the beauty of the ceremony – and of course the Wedding Kiss.[4]

So far of course this is all what might be called "vapor ware" or even more precisely the "vapor wedding package."

Orbital Sciences Corporation (OSC)
This Virginia-based company was NASA's choice in February 2008 to replace RpK for funding under their COTS program. Clearly there is more than a bit of irony here. This is because at one point OSC was the partner with Rocketplane-Kistler to develop access to low earth orbit, but Orbital pulled out over the design path that was being pursued.

In late 2007 / early 2008, OSC was selected from among several other finalists to win the NASA Space Act Agreement award after RpK's contract was canceled for not failing to meet financial milestones. OSC's award from NASA is reportedly worth $171-million. These funds are to subsidize the R&D needed to build and demonstrate a launch system capable of delivering cargo to the International Space Station. This COTS program is a gamble by NASA to stimulate privately-owned alternatives to replace the Space Shuttle for ferrying crew and cargo to the ISS. Since a single flight of the Space Shuttle is estimated to cost anywhere from $500-million to a billion dollars, this seems to many as a reasonable gamble to take.

It was a bright young man, not long out of Harvard Business School, named David Thompson who founded the Orbital Sciences Corporation (OSC) in 1982. In the last quarter century, OSC has grown from a small firm dabbling at the edges of the space industry to become a billion-dollar firm with a full palette of services and products. Today, the company builds a variety of civil and military satellites for telecommunications, remote sensing, surveillance and scientific missions and a series of small to medium size launch vehicles. Its Pegasus and Taurus vehicles now almost dominate the low end of the market and OSC plans to use their Taurus II booster as part of the COTS program. They have also announced that the initial launch will be from the Mid-Atlantic Regional Spaceport at Wallops Island.

OSC's experience with NASA has been a series of ups and downs – figuratively and literally. In August 1996, they took on the development of the pilotless X-34 experimental space plane under a 50-month contract awarded by NASA. This low cost demonstrator was to be dropped from a carrier jet vehicle and then fly up to Mach 6 speeds and to an altitude of 260,000 feet. This contract, like so many of NASA X-series of craft, was canceled in mid-stream in 2001 for a variety of reasons – the most important one likely being the huge demands represented by the operation of the Space Shuttle and the building of the International Space Station. The X-34 space plane still sits rather forlornly at the Orbital Science Corporation's headquarters site, but now, seven years later, serves as an inspiration for future possibilities.

The hope of both NASA COTS contractors – Space X and OSC – is that during the next decade they will not only be able to ferry astronauts to the ISS, but that there will also be a demand to carry passengers to commercial space habitats. The Genesis inflatable space habitats now launched by Bigelow Aerospace, suggests that this additional market could actually be there, at least someday.

XCOR (Xerus)

After many years of developing and producing safe, reliable rocket engines and rocket-powered vehicles, XCOR is now embarked on the development of a viable space plane for sub-orbital flights to support a space tourism business. In addition, the company is working in partnership with ATK / Alliant to develop a low-cost LOX / methane rocket propulsion for NASA's Crew Exploration Vehicle (CEV). ATK will use the engine as a basis for development of the final flight-weight hardware. These engines are being designed to return the CEV from Lunar orbit to Earth and to perform in-space manoeuvring.

Figure 3.4: The XCOR-Xerus sub-orbital vehicle for space tourism
(Photo courtesy of Space Adventures)

XCOR's president, Jeff Greason, said:

"This is a wonderful opportunity for NASA and for XCOR. NASA is reaching out to small businesses and this contract is an excellent example. Both private industry and the government will benefit from this project, as well as future users of space vehicles."[5]

The company has received a Reusable Launch Vehicle mission license from the FAA's Office of Commercial Space Transportation (AST). The license, which is the first for a commercial reusable launch vehicle (RLV) that is launched and recovered from the ground, will be used to test RLV technologies prior to sub-orbital passenger travel. The launch license does not yet cover passenger operations though it does allow for revenue payload flights after initial tests are completed. Testing is being carried out at the Mojave Space and Airport in Mojave, California.

This spaceport, as now licensed by the FAA, is the first inland launch facility to be authorized for commercial launches. According to XCOR's Government Liaison Officer, Randall Clague:

"This license covers the full flight test program conducted in a designated test area. A significant feature of the license is that it allows the pilot to do an incremental series of flight tests – without preplanning each trajectory."[6]

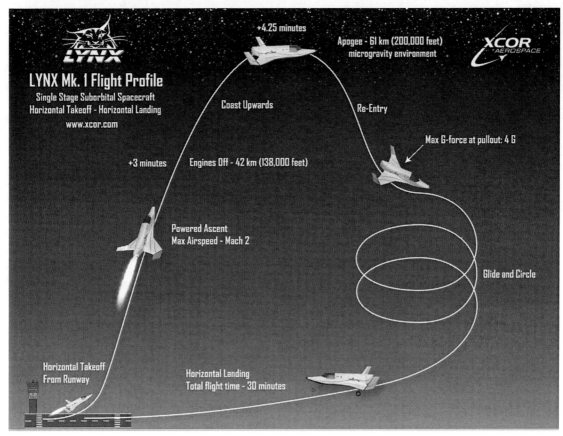

Figure 3.5: The 30-minute 'cut price' flight offered by XCOR
(Diagram courtesy of XCOR)

In March 2008, Greason, pulled a rabbit out of the ever expanding space tourism hat by announcing amid much fanfare his new "Lynx" spaceliner. This new Lynx craft would accommodate only a pilot and a single passenger who would ride in the copilot seat to an altitude of 37 miles or about 60 kilometers – far below the past definition of outer space. (In fact, astronauts are defined as having gone into space by going 62.5 miles or 100 kilometers into space.) However, Greason's "cut rate" version of a space-liner, with an estimated cost of $10-million (versus Burt Rutan's SpaceShipTwo at $50-million), is offering much lower passenger fares. His pseudo space ride, also planned for a 2010 start-up, is projected to cost a mere $100,000 compared with a ticket price around $250,000 for a flight on Virgin Galactic. Of course the passenger gets far less – a flight of about thirty minutes duration and at most two minutes of weightlessness – versus a two-hour flight and four minutes of weightlessness. Clearly the more successful entrants there are into the field the more flight costs will decline. In the case of XCOR and Greason, however, it is not only the costs that are projected to decline but the altitudes as well.

Greason is one of the more outspoken members of the Personal Spaceflight Federation but he is also one of those guys with the "vision thing." He not only expects the space plane and space tourism businesses to grow and thrive but also to evolve in new and interesting ways. His vision is that there will be vertical evolution where different firms will develop particular expertise and competence with regard to various subsystems. Thus some like XCOR will develop propulsion systems, others will design avionics systems, or escape modules, or space flight suits. This model, if it proves to be true, could allow a number of today's players to evolve into major components of an overall private space flight industry. Greason always talks a good game. Time will tell if his vision is the one that plays out.

SpaceDev-Dreamchaser

The Poway, California, based SpaceDev Corporation is designing a reusable, piloted, sub-orbital space ship known as Dreamchaser. This vehicle is envisioned as a launch system that could, in time, be scaled up to safely and economically transport passengers to and from low earth orbit. This upgraded vehicle could, in theory, be used to re-supply the International Space Station, even though it was not chosen by NASA for the COTS development program.

SpaceDev signed a Space Act Memorandum of Understanding (MOU) with the NASA Ames Research Center in 2005 with regard to this development project. This non-binding MOU confirms the intention of the two parties to explore novel, hybrid propulsion-based hypersonic test beds for routine human space access to low earth orbit. It allows both NASA and u to investigate, either independently or on a cooperative basis, the potential of using hybrid propulsion, as well as other technologies, to develop new, piloted small launch vehicles. One of these possible projects for mutual collaboration is the SpaceDev Dreamchaser™ project.[7]

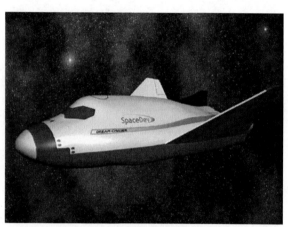

Figure 3.6: Artist's Representation of the
Dreamchaser Space plane Vehicle
(Graphic courtesy of SpaceDev)

As noted in the previous chapter, the rocket motors used in the SpaceShipOne X-Prize winning vehicle employs SpaceDev's patented hybrid propulsion system. This is the rocket motor that uses Nitrous Oxide (i.e. laughing gas) as the oxidizer and neoprene rubber as the propulsive fuel. Traditional rocket motors use two liquids, or a solid propellant that combines the fuel and oxidizer, but both types of rocket motors are highly explosive and therefore also highly dangerous. Further, all solid motors produce large quantities of toxic exhaust. SpaceDev's hybrid rocket motors do not detonate like solid or liquid rocket motors. This unique propulsion system not only produces a high performance thrust, but the nitrous oxide can be "turned off" to shut down the engine in the event of an emergency.

Unlike the design used in SpaceShipOne, the SpaceDev Dreamchaser™ would take-off vertically, like most launch vehicles, and then glide back for a normal horizontal runway landing.

In March 2008, SpaceDev announced the award of a contract by NASA's Marshall Space Flight Center for the research and development of its next generation proprietary annular hybrid rocket motors. The contract work will be performed over a six-month period, during which SpaceDev is expected to conduct development testing with the objective of validating performance parameters. This technology could provide significant improvements to hybrid rockets in terms of both performance and packaging efficiency that could significantly broaden their overall applicability and value.

Mark N. Sirangelo, SpaceDev's Chairman and Chief Executive Officer, said:

"We've been continuing our development of advanced rocket motors for spacecraft and this is a strong next step in that process. Our cooperation with NASA and specifically with the Marshall Space Flight Center is of major importance to us. Their experience and knowledge will provide considerable added value to the important work being conducted by our propulsion team. This technology could allow hybrid rocket motors to establish a presence in areas they have not been historically considered in commercial, civil space and military applications."

In August 2008, SpaceDev strengthened its Board of Directors with the appointment of Patricia Grace Smith, the former Associate Administrator of Commercial Space Transportation for the Federal Aviation Administration. She brings over 25 years of experience including 14 years with the Federal Communications Commission (FCC). At the FAA, she headed the office responsible for licensing, regulating and promoting the US commercial space transportation industry for 11 years.

The announcement by SpaceDev highlighted Ms. Smith's leadership at the Office of Commercial Space Transportation. It expressed the view that the FAA has become a recognized global leader in support of private human space flight. Ms. Smith has worked closely with the FAA's two lines of business to apply aviation expertise where appropriate to space issues and to address the impact of space flight on the National Airspace System.

"We are honored to have someone of Patti's experience join our board," said Mark N. Sirangelo, SpaceDev's Chairman and CEO. "As we continue to lead our industry in the commercialization of space, Patti's depth of experience and knowledge will be an important strategic addition to our Company."

This news was quickly followed by another announcement by the company of a multi-year contract with Scaled Composites to work on the development and testing of a production rocket motor for Space ShipTwo, as noted earlier in Chapter 2.

The Next Tier
There is a further group of contenders listed in Appendix A who think they can also succeed in the space tourism marketplace. Some of these companies, like the enterprises already covered, may also have their eye on a new executive supersonic transport business as well. Some of these hopefuls are comparatively small organizations; others (like Alliant) are part of a much larger group with far greater resources. The progress to date by some of these companies is briefly profiled here.

Kelly Space & Technology Inc.[8]
One of the more innovative concepts for achieving a low cost access to space is being pursued by the Kelly Space & Technology Inc. This privately-held company operates from the San Bernadino International

Figure 3.7: Reusable KST Elipse –Towed to 6 Kilometers Altitude Before Release
(Courtesy of Kelly Space & Technology Inc.)

Airport in California and is trying to develop a wide array of launch vehicles for both unmanned and manned missions to space. KST has developed and patented a so-called "tow launch" technology. The system employs a modified 747 airliner to tow the KST Eclipse Astroliner to 20,000- feet (about 6-kilometers) before separation. The Astroliner then accelerates to speeds of up to 6,000 miles per hour and an altitude of 112 kilometers (400,000 ft) where it releases the second stage rocket and satellite payload, which is then to be injected into low earth orbit. This system is designed to both reduce launch risk and cost per kilogram of payload.

KST is developing a wide range of spacecraft and launch systems to meet any number of needs including satellite launch systems, ISS re-supply, and crew and cargo transport vehicles. The company's web site simulation of how the KST launch system works is a beautiful thing to behold. Now for getting the real world machinery to do the same trick.

Alliant ATK Pathfinder

In 2007, Alliant Techsystems (ATK) announced the successful completion of the on-pad assembly of its Pathfinder-ALV vehicle. This involved the mating of the Pathfinder with the ATK-designed launch vehicle, the ALV X1. However, a test flight of this sub-orbital rocket in August 2008, carrying two NASA hypersonic flight experiments, was destroyed by range officials shortly after its launch from the US space agency's Wallops Flight Facility on Virginia's eastern shore.

The ALV X-1 program is part of ATK's plan to develop a low-cost launch vehicle for the operational responsive space (ORS) market for the US Department of Defense and other customers. Possible ORS program requirements include the delivery of small payloads to low-earth orbit in support of DOD missions – such as the deployment of a dedicated communications satellite to meet specific urgent battlefield requirements in an area lacking telecoms capacity. ORS requirements could also include NASA scientific missions, and commercial and university satellite programs.

ATK differs from many other start-up companies seeking to enter the space tourism market. It is part of a $3.4-billion advanced weapon and space systems company, perhaps most noted for its Thiokol Division that developed the solid motor rockets for the Space Shuttle. ATK, after its latest mergers and acquisitions, now employs approximately 15,000 people in 22 states within the US. Thus, ATK's business plan and its capabilities are quite different from most other organizations covered here.

The Vice President of Strategy and Business of the ATK Launch Services group, Charlie Precourt, announced back in October, 2006 that:

> "...the completion of our pathfinder operation builds on last year's successful sounding rocket flight test and brings us one step closer to realizing our goal of developing an affordable launch vehicle that supports numerous military, scientific and commercial missions. By incorporating off-the-shelf components and technologies into the program and minimizing ground support requirements, we believe ATK can offer our customers a reliable and low life-cycle cost launch vehicle."[9]

As well as its corporate size, Alliant is unlike many other organizations trying to develop space vehicles for the space tourism business by pursuing a different model. This is to upgrade existing proven rocket engines with a good deal of previous successful launch experience for the research package launch market (i.e. the US Department of Defense, NASA, Universities et al as customers) and then to transition to the "space tourism" market as opportunity and experience so allows.

Aera Space Tours – Altairis[10]

The various companies headed by entrepreneur William Sprague have had a number of different names over time. These have included Aera Space Tours, Altairis Rocket Company and Sprague Space Corporation. The Altairis rocket system is currently under development and reservations are being taken for sub-orbital flights with seats on the first launch being reserved against an initial fare of $250,000.

When he announced his plans in 2005, Sprague said the company was "on course" to start test flights in August 2006, conduct the first commercial tourist flight `during the following December and go into regular service shortly thereafter – a timetable which has clearly slipped. However, Aera has completed arrangements for launching its vehicle from Cape Canaveral, including flight logistics support.

Altairis is designed to launch six passenger astronauts and one mission commander into space and bring them safely back to Earth. All "astronauts" must meet certain minimum physical requirements, according to Bill Sprague who is president, CEO and chief scientist of the Temecula, Calif.-based company, which bills itself as 'a leader in the luxury space travel industry'. The Altairis program has promised much on its web site, and only time will tell what it can actually deliver from the launch pad.

JP Aerospace – The "Other Space Program"

JP Aerospace represents a strikingly different and novel approach to the space tourism and space research market. It has designed a viable concept for lighter-than-air systems to access Low Earth Orbit. The company is headquartered in Rancho Cordova, CA, but the location for its test program has been the West Texas Spaceport.

The company is currently developing new technology to fly a balloon (or perhaps more accurately an airship) to extremely high altitudes and then use ion engines to reach low earth orbit with a special ultra-light weight high altitude vehicle. The concept is as simple as using "the Earth's atmosphere as a ladder to space." This project was inspired by the early work of the scientist James Van Allen, who was one of the pioneers in balloon-launched rockets. JP Aerospace engineers have indicated that by using this technology, they might also reduce costs and increase safety in access to Low Earth Orbit. The advocates of this quite logical approach start with the common sense observation that balloons have carried people and machines to the edge of space for over seventy years. Therefore, with some creative engineering, balloons and highly efficient ion engines can accomplish the task with greater cost efficiency, less pollution and greater safety than approaches dependent on rockets using chemical propulsion systems.

For many years, flying an airship directly from the ground to orbit has not been considered practical for a number of technical and aerodynamic reasons. An airship large enough to reach orbit would not survive the winds near the surface of the Earth. Conversely, an airship that could fly from the ground to upper atmosphere would not be light enough to reach space. Thus, the need to reach low earth orbit with a viable lighter-than-air craft that is structurally sound and of proven reliability and safety

Figure 3.8: The JPA Ascender Balloon System under construction in its hanger
(Photo Courtesy of JP Aerospace)

requires an innovative and unique architecture, augmented by some additional propulsive force.

JP Aerospace claims to have overcome these constraints by the development of a very innovative three-part system. The first part is an atmospheric airship, the Ascender (not to be confused with other vehicles using this same name). It is designed to be capable of traveling from the surface of the Earth to an altitude of 42-kilometers (or 26-miles). The vehicle is operated by a crew of three and can be configured for cargo or passengers. It is a hybrid vehicle that uses a combination of buoyancy and aerodynamic lift to fly. It is reasonably maneuverable and driven by propellers designed to operate in a near vacuum.

The second part of the system is a sub-orbital space station in "proto-space" or "sub space." This is designed to be a permanent, crewed facility parked at 42- kilometers and is called a Dark Sky Station (DSS) – "the way station" to space. The DSS is envisioned as the destination of the Ascender, the departure port for the orbital airship, as a proto-space habitat for tourists and as a research station. Initially, the DSS will also be the construction facility for the orbital vehicle.

The third part of the system is the airship / dynamic vehicle that flies directly to orbit. To utilize the few molecules per cubic meter of gas at extreme altitudes, this craft must of necessity be of a large volume.

The initial test vehicle design, therefore, is 1.82-kilometers (or 6,000-feet long.) The reduced force and speeds of the wind at these super altitudes allow a craft of much greater fragility than the Ascender designed for lower altitudes. This very large and high volume airship uses its great buoyancy to climb to 60.6-kilometers (200,000-feet), using electric ion propulsion to gradually accelerate. As it accelerates, it dynamically climbs and over several days it reaches orbital velocity. This approach appears to offer a number of advantages:

- All three parts of the ascension system are either long term or reusable in nature (with the possibility of several missions before servicing).
- The system is scaleable to handle larger scale payloads.
- Large structures can be placed already assembled in orbit using this type of ascent system
- The ascension system provides a safe and reliable way to access space
- Low cost access to space for significant payloads.
- Avoids large thermal gradients for re-entry
- There are no large fuel tanks or rocket motors to explode
- It is a "green" or "clean" system that is much more environmentally friendly than a conventional space plane.

Once in orbit, the airship truly becomes a spacecraft. With solar / electric propulsion, this type of vehicle could in theory proceed to multiple locations in the solar system such as the Moon or Mars.

The Ascender and Dark Sky Station project is now over two decades in development. There have been over eighty real hardware test flights and countless development tests. One of the remarkable aspects of this project is that it is being built essentially with existing technology. However, new polyimide, kapton or other materials and the latest in ion engine, solar cell and fuel cell technology could be applied to advance the design and performance further. A prototype high altitude airship has been built and is awaiting test flights and the first crewed DSS is scheduled to fly in 2008.

Remarkably, this new way to space has not required huge capital expenditures to accomplish; each component has its own business application and funding source. Thus the JP Aerospace development program operates as a pay-as-you-go system. Clients seeking an improved high-altitude reconnaissance vehicle provided funding for the atmospheric airship. The DSS development, on the other hand, has been supported by multiple customers in the telecommunications community. JP Aerospace estimates that it is seven years from completion of the entire system.

The Transformational Space Corporation (t/Space)

This company is based in Reston, Virginia. It was formed in 2004 to respond to NASA's plans to implement the President's Vision for Space Exploration. The company was one of eight winners in NASA's "Concept Exploration and Refinement" competition to advise the agency on the best architecture for Moon-Mars exploration and the best initial design for the Crew Exploration Vehicle (CEV). The effort kicked off in August 2004 with a $3-million contract that was extended in March 2005 with another $3-million. The company, however, was an unsuccessful finalist in the $500-million NASA COTS program to create low-cost cargo and passenger vehicles.

t/Space's Crew Transfer Vehicle (CXV) was designed to carry crew to the International Space Station or other low Earth orbit destinations in response to the NASA solicitation. It is currently being marketed as a way to offer commercial access to low earth orbit, but on the basis of a relatively high cost of $20-million per flight. The CXV's design is based on the Discoverer and Corona capsules that have demonstrated some 400 successful missions. Thus, the CXV utilizes a proven re-entry design in order to maximize safety and survivability of the crew. The CXV is launched into space on a booster released from a cargo carrier aircraft, which enhances safety, performance, and flexibility.

The t/Space web site notes that the design philosophy behind the CXV is driven by "Simplicity, Survivability, and Affordability." These principles have guided both the design of the CXV and the Earth-to-orbit launch system that is being developed by Air Launch, in conjunction with the Falcon launch vehicle system that is planned as the second stage. According to t/Space, the design has been optimized to take people into orbit safely and at a price low enough to enable development of a new market for personal space flight for people who wish to go beyond sub-orbital flights and actually go into low earth orbit for multiple orbits of the Earth. The cost of the t/Space system, if actually successfully implemented, would in theory be at least three times less than the cost NASA is currently paying to the Russian Federated Space Agency for a Soyuz launch to the ISS.

Safety considerations in the t/Space design include air launching from an altitude of 25,000 to 30,000 ft. that would provide far safer abort modes during the first few critical seconds of flight. Also, the booster utilizes a simple design with fewer failure modes. The engine design has very few moving parts – there are no turbo-pumps or pressurization systems. The capsule shape provides safe re-entry, even if all control systems have failed, since it will automatically right itself as it descends into the atmosphere regardless of its initial orientation. The capsule's thermal protection system is a double layer of SIRCA tiles, developed by NASA Ames Research Center, which shields the vehicle and crew from the heat of reentry. Water landing of the capsule uses proven systems from the Apollo program.

The t/Space Earth-to-orbit system has been designed from the ground up to reduce the cost of access to space for NASA astronauts. The current cost "per astronaut" that NASA is paying for a US seat on the Russian Soyuz system is $20-million. The CXV system, in theory, would entail a round trip cost to Low Earth Orbit and return of about $7-million.

The Third Tier: "The Wannabes"
There are still other companies that have entered into the private space business in recent years. These are very hopeful enterprises that nevertheless are generally considered to be "longer odds wannabes." These companies include the following:

Advent Launch Services was formed in Houston, Texas, in 1999, but the Advent concept started long before that. As part of the Future Programs Office at NASA's Johnson Space Center, several individuals were studying ways to make the Shuttle program more cost-efficient. At the time (1990), NASA was considering boosters using liquid propellants to replace the solid propellant boosters for the Shuttle system. The initial cost analysis of the Shuttle system prompted the team's engineers to consider features for a launcher concept that could achieve minimum cost.

The current Advent concept is a spacecraft that launches vertically from water and lands horizontally like a seaplane. It is a winged rocket designed to glide down to the ocean surface for a safe, controlled landing. It would, in theory, be safer than an airliner because it is mechanically much simpler, having fewer components, a shorter run time and very robust mechanical parts. In fact, the Advent vehicle requires only seven simple "on" or "off" signals to fully control the propulsion system. The guidance system uses redundant control surfaces on the trailing edges of the wing for propulsion efficiency and steering, atmospheric re-entry and aerodynamic control for gliding and landing.[11]

Air Boss Aerospace has been specializing in high-tech aerospace programs for over 16 years and is headquartered in Colorado Springs, Colorado. It is working on the development of proof-of-concept and prototype aircraft, ranging from low performance general aviation aircraft and UAVs (unmanned aerial vehicles) to business jet class vehicles and spacecraft.

Most notably, Air Boss developed the first aircraft to start the VLJ movement in the business jet marketplace. They did so, when a suitable jet engine did not exist, by taking an existing turbine and redesigning it to fit the aircraft needs. Currently, AirBoss is a partner in the development of two spacecraft. Both vehicles have the capability of horizontal landing, and one is powered by a regenerative methane rocket motor.[12]

Air Boss is also working with C&Space Industries (CSI) of the Republic of Korea on the development of a sub-orbital space plane called Proteus. This project is described in greater detail in Chapter 4.

DTI Associates of Arlington, Virginia, markets and offers integration services for the Terrier-Orion after purchasing all intellectual property rights to the rocket from SPACEHAB in July 2001. A more powerful version of the Terrier-Orion rocket uses the Terrier Mk 70 motor as its first stage. This version was used for two FAA / AST-licensed sub-orbital launches performed by Astrotech Space Operations / DTI at the Woomera Instrumented Range in Australia in 2001 and 2002. The second flight, in July 2002, successfully flew the HyShot scramjet engine experiment. The current status of DTI Associates operations is not clearly known.

Excalibur Almaz is a private space flight company which plans to orbit manned spacecraft, by using modernized TKS space capsules and Almaz space stations, derived from the formerly secret Soviet space program. Missions will support orbital space tourism, and provide test beds for experiments in a microgravity environment. The group (based in Douglas, Isle of Man, with offices in Houston and Moscow) is following "a lightweight and efficient business mode," by owning its spacecraft but contracting expert services, including refurbishment, launch, control, and recovery.

Company founders include: CEO and space law expert Art Dula, CFO and space commercialization veteran Buckner Hightower, and Sales & Marketing Vice President Chris Stott. Stott is also CEO of ManSat and is on the board of the International Space University. The company's COO is US Air Force General (ret.) Dirk Jameson, who once commanded the Air Force's Vandenberg missile launching base. Chief of spacecraft operations is Leroy Chiao, formerly a NASA astronaut and Commander of the International Space Station.

An Advisory Board includes former Johnson Space Center Director, George Abbey; former Kennedy Space Center Director and former President of Lockheed Martin Space Operations, Jay Honeycutt; former space shuttle astronaut and VASIMR plasma rocket engine inventor, Franklin Chang-Diaz; former European Space Agency astronaut Jean-Loup Chrétien; and former Russian cosmonauts, Vladimir Titov and Yuri Glaskov.

The TKS-derived space capsules resemble American Gemini capsules, but unlike the two-person Gemini, they are reusable, and can carry three passengers or operate autonomously. They can launch atop any of several rockets of various spacefaring countries, and they possess a Launch Escape System to ensure the safety of their passengers. They use parachutes and retrorockets to return to earth on land or on water. The Almaz-derived space stations are closely related to modules used on the International Space Station, and on the Soviet and Russian Salyut and Mir space stations. Excalibur Almaz's space stations will feature the largest windows ever on spacecraft.

HARC. The Huntsville-based High Altitude Research Corporation (HARC) originally announced plans to develop a sub-orbital vehicle, the Liberator, to compete for the X Prize in 2003 and gave a select audience its first detailed look at the vehicle, showing off hardware already constructed. They planned to fly the Liberator into space by late in 2008. Further details are not currently available, but based on licensing and testing, their objectives seem unlikely.

Interorbital Systems Corporation (IOS) is an American aerospace company based in Mojave, California. It was founded in 1996 by Roderick and Randa Millirond, who also co-founded Trans Lunar Research. The company is currently working on a line of launch vehicles aimed at winning the newly created America's Space Prize. Interorbital Systems was also a competitor in the Ansari X-Prize. The company has successfully tested a number of rocket engines in the 500-5,000 foot pounds (i.e. 2 to 22 kiloNewtons) thrust range.

InterOrbital Systems' prototypes are all designed to be amphibious (capable of launch from land or sea). Although IOS currently launches its spacecraft from Mojave Space Port, it plans to develop and implement a method for launching directly from the water without a launch platform. The vehicle is towed into position by a 33m (110 ft) crew boat, then a ballast is suspended from the submerged tail-section to provide proper orientation and stability during launch. According to IOS, sea launches are inherently less dangerous than land launches because they reduce the chance of damage to people or property in the event of a failed launch. As a result it is easier and less expensive to obtain a license for sea operations.

In July 2007, IOS announced that Tim Reed, a Mid-western businessman and adventure traveler, is the first to purchase a ticket for a week-long orbital expedition aboard the five-passenger IOS Neptune Spaceliner, scheduled for launch in 2008. The company claims that the ocean-launched Neptune will be the first manned orbital launch vehicle built totally without government funding. Reed bought the first of Interorbital's $250,000 "promotional fare" spaceline tickets, which come with a complete rebate of the purchase price, redeemable two years after the crew member's flight. Only nine of the reduced fare tickets remain for sale, with additional purchases pending. After these advance-purchase tickets are sold, the price for an orbital expedition with IOS will revert to the standard fare of $2 million per passenger.[13]

Crew members will undergo a rigorous thirty-day training program before the sea-based launch out of the Pacific Ocean off the coast of California. The novel design of the Neptune, a SAAHTO (stage-and-a-half-to-orbit) vehicle, allows the rocket's two spherical 21-foot diameter liquid oxygen tanks to serve as spacious habitat for the five-person crew during their seven-day stay in orbit. A capsule return with ocean splashdown and recovery completes their space adventure. Neptune is one of 4 prototype vehicles being developed by IOS. The others are:

- Neutrino, the smallest spacecraft is designed for suborbital flight.
- Tachyon designed for orbital flight.
- Sea Star NSLV designed to reach orbit and with a launch planned for 2007 from the Pacific Ocean near Los Angeles. (Interorbital incorrectly claims that if Sea Star NSLV is launched on schedule it will be the world's first private satellite-launching rocket, a title long held by Orbital Sciences Corporation's Pegasus launch vehicle.)
- Lorrey Aerospace is a project headed by Mike Lorrey of New Hampshire. Lorrey Aerospace's business plane is based on the conversion of an F-106 Dart high altitude stock jet into a higher performance X 106 Dart ramjet that can flight into low earth orbit with a pilot, passenger and 500 pounds of cargo. This is a small, low capitalized project that has not updated its website for some months.[14]
- Neutrino, the smallest spacecraft is designed for suborbital flight.
- Tachyon designed for orbital flight.
- Sea Star NSLV designed to reach orbit and with a launch planned for 2007 from the Pacific Ocean near Los Angeles. (Interorbital incorrectly claims that if Sea Star NSLV is launched on schedule it will be the world's first private satellite-launching rocket, a title long held by Orbital Sciences Corporation's Pegasus launch vehicle.)
- Lorrey Aerospace is a project headed by Mike Lorrey of New Hampshire. Lorrey Aerospace's business plane is based on the conversion of an F-106 Dart high altitude stock jet into a higher performance X 106 Dart ramjet that can flight into low earth orbit with a pilot, passenger and 500 pounds of cargo. This is a small, low capitalized project that has not updated its website for some months.[14]

Masten Space Systems is a Mojave, California, based rocket company that relocated from the higher rent district of Santa Clara, California. Masten Space Systems Inc. is currently developing a line of reusable Vertical Takeoff and Vertical Landing (VTVL) spacecraft, and related rocket propulsion hardware. They have announced their intention to continue to compete against Armadillo Aerospace in the Lunar Lander Analog Challenge as sponsored by NASA. (Their crude prototype actually came in a distant second to Armadillo Aerospace. Their apparatus of Rube Goldberg-like complexity looked more like a lab mockup that a flight craft.

Their XA (eXtreme Altitude) line of suborbital (VTVL) vehicles represent a family of incrementally developed vehicles starting with technology demonstrators, and leading to commercially operated manned and unmanned suborbital launch vehicles. The XA series of suborbital vehicles will also help lay the groundwork and prove out some of the technologies necessary for future TSTO orbital launch vehicles, lunar landers, and other spacecraft.

The first vehicle in their line of VTOL spacecraft is the XA-0.1. It will serve as a technology demonstrator, testing out engines, controls, and other systems for future vehicles. XA-0.1 uses four Liquid Oxygen / Isopropyl Alcohol rocket engines, each of which produce up to 500lb of thrust at full throttle. The initial flight of the XA-0.1 vehicle is planned to go shortly. The colorful slogan for the Masten Space Systems is "Just Gas 'em and Go!" So far they have a long way to go.[15]

Space Access LLC is a two stage to orbit spaceplane design. The first stage uses an "ejector ramjet" to power a hybrid air breathing engine in the first stage to reduce aerodynamic drag. The second stage would have one of several different configurations. All of the designs envision that the second stage would be contained inside the first stage. This second stage would be ejected from the first stage at very high altitude as indicated in the figure below.

Space Transport Corporation, or STC, is a company based in a rural part of the US state of Washington. STC's goal is quite simply to commercialize space. The company was founded in 2002 by Eric Meier and

Philip Storm. The company plans to provide small payload launch and space tourism services. This quite small enterprise was founded in response to the X-Prize. It has undertaken the development of the Rubicon I and II vehicles, but after launch failures and the claiming of the Ansari X Prize by SpaceShipOne, the launch systems have been down-sized to the Spartan and N-Solv project that only can launch 5 Kg to 10 Kg in payload. If these can be proved in launch tests, then larger scale rockets are planned. Last known capitalization was in the $100,000 or so range.[16]

Star Systems Inc. was founded Star Systems Inc. in Phoenix, Arizona in 1993 by Morris Jarvis and, with group of fellow-enthusiasts, he began building the prototype of his Hermes spacecraft in his garage in 2003. The result went on public display for the first time at the Intel Developer Forum in August 2008. Morris and his team are building Hermes out of their own pockets and say they need about $1.5 million to finish the test work and begin regular space flights.

Jarvis is sponsored by his employer, Intel and by other companies including ADI Engineering, Dot Hill, GE Fanuc, MicroSun, and National Instruments. The first flight of Hermes was scheduled for October 2008 – an unmanned, tethered flight at Utah's Bonneville Salt Flats during which Jarvis will control the spacecraft from a remote cockpit on the ground. Once the tethered flights are completed, Hermes will be equipped with an engine pod so it can be flown like an airplane, unattached to the ground. In the next stage, Hermes will be towed to about 113,000 feet by an ultra-high altitude helium balloon, then flown back to earth using the remote cockpit. Finally, another test will be undertaken with an onboard pilot. If all goes well, Jarvis and his colleagues plan to build a production version of Hermes and begin offering regular space flights thereafter if all goes well.

"Hermes is built on the premise that anyone who wants to should be able to take a trip into space," says Jarvis. "We hope to provide trips for about the price of a new car."[17] Of course a new Rolls Royce runs more than a quarter million dollars. We believe that he is shooting for something more like a Toyota Camry.

TGV Rocket was founded in April 1997, in the state of Maryland by Patrick Bahn. He developed the company with the founding management team of Dr. Earl Renaud and Kent Ewing. TGV received an initial design study contract for a reusable launch vehicle with the US Naval Research Laboratory, Washington, DC in 2003 and has expanded this line of business over the last 3 years. The company has serviced this work successfully from its Norman, Oklahoma location and has grown revenues to over $5-million in services. In 2004, TGV was awarded a small disadvantaged business 8(a) certification by the Small Business Administration (SBA). Patrick Bahn, company CEO, is still answering his own phone and his enterprise is very much a cash-strapped venture. TGV's pockets are far from deep although its ambitions remain high.[18]

Triton Systems is based in Houston, Texas and stems from the combination of several individual consulting operations that begun in the Houston area in the mid-1990s. Triton Systems, LLC was incorporated in Texas in July 2004 and in August 2005, opened its Houston office a few miles from the Johnson Space Center. The team consists of aerospace specialists in flight mechanics, trajectory optimization, computer software, aerodynamics and other related disciplines. It admittedly is taking an unconventional approach to the space tourism business using "underground thinking" to achieve low cost but reliable access to space.

Triton's president and chairman, Wes Kelly, is an air force veteran and aerospace engineer of three decades experience in industry. The company's Stellar-J launch vehicle has been described in several public forums. As an invited speaker, Wes Kelly provided accounts of Stellar J's progress at the 2003 and 2005 Space Access Society conferences in Phoenix, Arizona. Also in 2005, the AIAA Houston bi-monthly publication Horizons included "The Stellar-J Launch Vehicle" as a feature and cover article along with editorial comment. Although the Stellar-J Launch vehicle remains a "long shot," its design features may prove valuable to other projects.[19]

UP Aerospace of Hartford, Conn. describes itself as "the world's premier supplier of low-cost space access." It recently launched its unmanned SpaceLoft XL vehicle from the White Sands Missile Range and is scheduling up to 30 space launches per year from New Mexico's "Spaceport America."

The SpaceLoft XL vehicle can launch up to 110 pounds (50 kilograms) of scientific, educational, and entrepreneurial payloads into space, with an altitude capability of up to 140 miles (225 kilometers). It gained a high degree of notoriety with its April 28, 2007 launch of the ashes remains of Star Trek James

"Scotty" Doohan, and Mercury Astronaut L. Gordon Cooper along with a number of others. The SpaceLoft XL is not a human rated launch capability and current business plans apparently are aimed primarily at launching only educational and scientific payloads. Since its payload capacity is limited to (25 cm x 2.2 meters) (i.e. 10 inches x 7 feet) human payload would be more than a bit difficult for most passengers. Longer term plans for a manned capability are nevertheless likely still being considered.[20]

Vela Technology Development is working on the SPACE CRUISER, a space plane project that is designed to carry six or more passengers. It is reported that the cost would be $50,000 per seat for a trip into sub-orbital space.

The design is a business-jet like design that would be air-launched from a large supersonic carrier aircraft. This design could evolve to support supersonic business transport and, as an evolutionary development, it could result in a lower cost space tourism business.[21]

Wickman Spacecraft and Propulsion is now celebrating its 25th year in business. Founded in 1981 in Casper, Wyoming by John Wickman, the company supported Thiokol on the post-Challenger SRB field and nozzle joint redesign and United Technologies Chemical Systems division on the Titan recovery program. In the late 1980s, the founder teamed with Dr. Adolf Oberth to move the company into research and development.

In the 1990s, the mission of the company was expanded to include manufacturing. This goal led WSPC to explore new areas of propulsion, launch vehicle construction and lunar base design and construction. Building on its previous work, WSPC has pushed the boundaries of PSAN solid rocket propellant technology beyond the dreams of earlier rocket pioneers. WSPC was the first company to develop the technology for using low cost commercial grade materials in solid rocket motors and launch vehicles. The company claims that its "cutting edge technologies" are now enabling it to achieve low cost access to space with an innovative Small Launch Vehicle and sounding rockets.

New engines will be needed that can burn materials found on the moon, Mars and outer planets. WSPC has invented the rocket propulsion technology of mixing metal powders found on the moon into liquid oxygen and making a successful rocket engine using it. This knowledge enabled WSPC to successfully develop a rocket engine that burns carbon dioxide and magnesium powder for use on future Mars missions. WSPC has built and tested a turbojet engine that will eventually be able to operate on Mars for durations of 30 minutes or perhaps longer. In addition to its small launch vehicle launch capabilities, Wickmans Space and Propulsion Corporation is also designing a high altitude Sharp Spaceplane for the Air Force.[22]

Conclusions

The great diversity of design and engineering concepts reflected by all of the above companies is a great strength for the emerging space tourism and commercial space market. But there is also danger in this diversity of approach. This new and emerging industry is to an extent only as strong as its weakest link. Small start up enterprises with slogans like: "Gas 'em Up and Go" suggests a level of development, technological sophistication, and the ability to provide safe mass access to sub-orbital space that simply does not exist. Such safe and low cost access will not be achieved for many years to come.

In short, false expectations and false confidence in our ability to achieve mass travel to space is a real danger. This heady optimism will crash with the first fatalities that emerge from commercial space industrial development. Overselling the safety, level of development and maturity of this new and fragile space tourism business must be avoided.

Effective regulatory oversight and careful legislative guidelines are key if this new enterprise is to succeed. FAA Associate Administrator, Patricia Grace Smith, who heads the Office of Space Commercialization, keeps saying that major risks are out there and cannot be ignored. "Licensing of space planes is not anything we will be able to do soon," she says.[23] She also notes that the first thing they ask is for consumers to read and sign the waiver statement that space tourism flights are "dangerous."

Clearly the various space plane companies represent a heady stew of creativity that helps to expose innovative ways to access space. Entrepreneurial skills and competitive designs will certainly drive down costs over time. Competing designs and companies not only lowers costs and prices but creates a "creativity crucible" that ultimately highlights the best ideas. If we can survive the next ten years of space plane shakeout without a major disaster, the chances are that a space tourism industry can truly emerge.

References:

1. Space Planes and Space Tourism – GW University, SACRI, 2007.
2. RpK Wins $207 Million Award to Demonstrate Space Station Servicing http://www.astroexpo.com/news/ August 21, 2006
3. Kelley Chambers, *The Journal Record*, December 26, 2007
4. Rocketplane Global press release, July 7 2008.
5. Marc Kaufman, "Another Firm Joins the Commercial Space Race," Washington Post, March 30, 2008, p. A-3 and http://www.spacenewsfeed.co.uk/2005/17April2005_19.html
6. XCOR press release, April 23 2004.
7. SpaceDev web site – http://www.spacedev.com/
8. Kelly Space & Technology web site – www.kellyspace.com
9. Alliant Techsystems, Inc. Press Release: October 10, 2006
10. Altairis – http://www.pythom.com/space flightpricewarsMay192005.shmtl
11. Advent Launch Services web site – http://www.adventlaunchservices.com/our_concept.html
12. Airboss Aerospace web site – www.airboss-aerospace.com.
13. Parade – June 25 2005 – http://archive.parade.com/2005/0626/0626_ride_in_space.html
14. Lorrey Aerospace web site – http://www.lorrey.biz/
15. Masten Space Systems web site – www.masten-space.com
16. Space Transport Corporation, Wikipedia – http://en.wikipedia.org/wiki/Space_Transport_Corporation
17. www.hermesspace.com.
18. TGV Rockets Inc. web site – http://www.tgv-rockets.com/
19. Triton Systems web site – http://www.stellar-j.com/
20. UP Aerospace web site – http://www.upaerospace.com/
21. Vela Technology – http://www.abo.fi/~mlindroo/spacemarkets/sld008.htm
22. Wickman Spacecraft & Propulsion Co. – http://www.space-rockets.com/sharp.html
23. Interview at Center for Strategic and International Studies (CSIS), Washington DC in April 2007

— Chapter 4 —

"There have been no final decisions on the matter, but on the horizon, we think ultimately there will be Europeans, Russians, and Japanese working on the Klipper reusable space vehicle." Daniel Sacotte, ESA director of the Human Space flight, Microgravity and Exploration program.
"Indian space officials say that their reusable vehicle will be superior to the shuttle because it will have a number of enhanced safety features and be much more cost effective." Leonard David, Space.com

THE INTERNATIONAL SCENE

The space tourism business is rapidly expanding. It is on the verge of becoming a true global enterprise. In ten years spaceports and space vehicle programs will circle the globe. Many of these new space tourism initiatives are commercial and entrepreneurial in nature. But in many countries, including China, India, France, Germany and Russia, there are projects backed by governmental agencies or even military-funded projects. Even in countries like France and Germany, where a private aerospace company is taking the lead, the national space agencies (i.e. DLR of Germany and CNES of France) are providing significant support.

In the US, there is generally great enthusiasm for commercial space initiatives. Private ambitions and heroic individual efforts are clearly in vogue. There is a historical feeling in America, remaining from past centuries and dating back to the settling of the Old West, that creative enterprise wins the day. This is sometimes simply called "Yankee Ingenuity." American's tend to believe in rags-to-riches entrepreneurs like the famous "Horatio Alger stories" where a poor boy from the slums can rise to great wealth and power. American literature is dotted by many heroes whose hard work and creativity let them rise to fame and fortune. Today's real life examples like Jeff Bezos, Paul Allen, Bill Gates, John Carmack, and Elon Musk are thus seen as "prototypical" Americans, destined to succeed against the odds.

Ever since the exciting flight of SpaceShipOne, the American public has been rooting for the private aerospace designers. The average man or woman in the street is likely to believe that these daring and entrepreneurial innovators will be able to produce a viable space plane better, faster and cheaper than a NASA.

But in other parts of the world the predominant vision is dramatically different. The deregulation and privatization of traditionally state-run activities elsewhere has for the most part been much slower than in the US And in this environment, it is not too surprising that the citizenry in many countries tends to believe that governments should undertake such advanced space and high tech research missions; private space initiatives still seem like Buck Rogers science fiction. This is still true in most of Europe.

Russia is a unique case unto itself. Here, the changes after the break-up of the rigidly-controlled and totalitarian Soviet Union created a headlong rush toward capital markets where and when available. The most agile of the new entrepreneurs found ways to achieve privatization within the structure of formerly state-owned assets. This has been particularly true of the technologically sophisticated but under capitalized space launcher industry.

In China and India, the government strictly oversees and controls the aerospace industry and in the Republic of Korea there is a close symbiotic relationship between the public and private sectors.

But public or private, an array of space planes, spaceports and other facilities for testing and training is mushrooming around the world. The game is afoot in Russia and in Europe – including at least France, Germany, Switzerland and the U.K. There are also new ambitious space programs in Canada, India, China, Australia, Israel, the Republic of Korea and even Argentina. Efforts have also been started in such unlikely locations as Romania and the Caribbean. Spaceport initiatives can be found in other locations such as Singapore and the Middle East. Although fewer than half of these various initiatives are strictly private commercial space enterprises, private money and commercial intellectual capital lies behind many of them. Each country's space initiative represents a unique story, with a varied but always interesting cast of characters.

Europe Plays Catch-up

In Europe there is now a pro-active relationship between private space ventures and the various official space agencies. All of the players are seeking to promote the rapid development of this new industry. In terms of technology, regulatory frameworks, testing, training and simulation facilities and spaceports, Europe is currently behind the US in the space tourism enterprise, but sincerely trying to play catch up. (Virgin Galactic, Sir Richard Branson's enterprise, of course, is an entirely different species – neither American nor European – and perhaps, as such, the leader of the pack.)

Only a decade ago topics such as space tourism and private space ventures were considered to be science fiction in Europe – somewhere on the outer fringe of thinking. Today this activity is seen as more and more realistic and achievable. The winning of the Ansari X-Prize in 2004 was a great boost to space tourism efforts in the US, but in Europe it strongly suggested they had made a strategic "boo boo" by not providing greater governmental support for the private sector. More than one political leader in Europe has asked whether the lackadaisical approach to space tourism over the previous decade might have been another muffed opportunity on the high-tech frontier of new commercial opportunities.

In 1996, for example, the faculty at the International Space University in Strasburg overruled a proposal by students and ISU Founder Peter Diamandis to undertake its final class project on this topic. But the European faculty, at that time, saw space tourism and space commercialization as too outré for a serious academic research project.

However, by 2004 the European Space Agency (ESA) had initiated a number of studies that showed the potential for developing the concept of commercial human space. Then, in July 2006, they announced a program involving private companies working on the development of crewed space vehicles for the space tourism market. This initiative by ESA was seen as a two-way process that will result in new and interesting inputs into its own launch technology program. European space officials have expressed the objective of establishing closer and on-going links between ESA and the emerging new space tourism industry.[1]

ESA is exploring what types of launch systems it will need to deploy to follow on from the successful French-led Ariane-5 launch vehicle. However, in some ways this multi-national agency is behind the national development programs of space planes that have already been underway for some time in the U.K., France and Germany. ESA has now created the FLPP (Future Launch Preparation Program) and is trying, in part, to leverage off development work already underway. In particular, the initiative by ESA is seeking to build on the Phoenix and Pre-X space plane developments by the EADS Space Transportation Corporation of Toulouse, France. It is also involved as a sponsor of Germany's Enterprise Program being undertaken by the TALIS Institute and DLR – the German Space Agency. In this case, the German development is also working in tandem with the Swiss Propulsion Lab. Both of these efforts are aimed at developing a Reusable Launch Vehicle (RLV) in the form of a space plane.

ESA is also undertaking a joint development launch program with Russia's Roskosmos agency for the Klipper re-usable launch system and the Advanced Crew Transportation System (ACTS). Other ESA activities in this field include a heavy-lift launch vehicle design by The Aerospace Institute of the Technical University of Berlin. In addition, there is a plan by the German Aerospace Center to develop a two-stage-to-orbit vehicle known as the DSL. Finally ESA is keeping an eye on work by Asociatia Romana pentru Cosmonautica si Aeronautica in Romania to develop the Orizont manned sub-orbital single stage to orbit (SSTO); and also work on various space plane designs by STAAR Research of Scotland.[2]

ESA is not just pursuing space plane technology with its various national partners. They are also exploring the market for space tourism from a business perspective and have undertaken a significant project to help private companies get their space tourism ventures off the ground. The latest initiative is called, in the snappy fashion of space agencies: "The Survey of European Privately-funded Vehicles for Commercial Human Space flight." The stated purpose of this ESA study is to assess the feasibility of various businesses' mission concepts. A number of private companies already involved in planning for space tourism are being invited to submit their plans to ESA where experts in the General Studies Program will down-select to a final three for further study. Each selected company is to receive 150,000-Euros to develop their plans further. In the past this type of hybrid, public-private undertaking would be considered to be very European in nature. But NASA's Commercial Orbital Transport System (COTS) – funded to the tune of nearly a half billion dollars in research effort – trumps the European measure by a wide margin.

The survey will critically review the spacecraft designs and mission profiles, ensuring they are technically and financially feasible. Another key element in the ESA program is an assessment of safety standards and requirements that are seen as necessary to make space tourism viable.

One of the many ESA projects aimed at the commercial space market has been the development of design concepts not only for space tourism flights, but for the much bigger hypersonic transport market. ESA has funded research with Reaction Engines Ltd. of Oxfordshire, England to develop designs for the so-called A2 hypersonic jet that might be able to fly at speeds of Mach 5 and carry up to 300 passengers between Europe and Australia in under 5 hours. This design is still at the conceptual stage and it is estimated that it might be as long as 25 years away from commercial market.

Alan Bond, a senior engineer and managing director at the company, said:

> "The A2 is designed to leave Brussels international airport, fly quietly and subsonically out into the north Atlantic at mach 0.9 before reaching mach 5 across the North Pole and heading over the Pacific to Australia. The flight time from Brussels to Australia, allowing for air traffic control, would be four hours 40 minutes. It sounds incredible by today's standards but I don't see why future generations can't make day trips to Australasia."[3]

One of the most significant parts of the design envisions a hydrogen-fueled engine that would avoid either nitrogen oxides or carbon oxides being spewed into the fragile upper atmosphere. The other key element here is the increasing interest in developing hypersonic transport craft to move people around the globe rather than just a quick jaunt up to peek into space.

eanwhile, under its FLPP program, ESA is seeking to build on the national initiatives already under way, especially in France and Germany. The focus is on the development of technologies needed for the next generation of launchers – expendable, reusable or somewhere in-between. One of the particular challenges is satisfying the stringent economic requirements of the space tourism market that requires vehicles that are not only safe and reliable, but also cost effective.

The Projects of EADS SPACE Transportation

The Toulouse, France, based EADS Space Transportation has been carrying out serious studies and new technology demonstrators for a space plane that could, at a later stage, be incorporated into the overall FLPP program. Initial flight tests of its Phoenix project were carried out in 2004. These tests were designed to explore the final approach and landing of a reusable launch vehicle (RLV). Another demonstrator, Pre-X, is the intermediate experimental atmospheric re-entry vehicle to be developed as part of FLPP. Pre-X successfully passed its System Requirement Review at the end of 2005.[4]

Figure 4.1: The Astrium Spaceplane – Under Development by EADS
(Courtesy of EADS SPACE Transportation)

EADS (including its Astrium subsidiary) is the largest developer of satellites and rockets in Europe and also owns the AirBus company. It is a largely Franco-German group but it has plants across Europe and in Britain as well. As such it is likely to be the dominant European entry into this high risk and capital intensive commercial spacecraft business.

Early in 2007, François Auque, the executive president of EADS Astrium, gave a first glimpse of possible ambitions. He stated:

> "Space tourism is already emerging through suborbital flights, and could evolve towards orbital flights if the initial business development is successful … These flights can evolve in parallel with exploration … private sponsored contests might be organised to undertake such missions as planetary rover races, solar sail races and the like."[5]

Then, at the Paris Air Show in June 2007, Astrium unveiled its revolutionary new vehicle for space tourism. This spacecraft, the size of a business jet, is designed to carry four passengers 100-km up into space giving more than three minutes of "zero G" or weightlessness. Guests at the Paris event were shown a full sized mock-up of the forward section of the revolutionary craft including its Marc Newson designed cabin.[6]

According to Astrium, the passenger-carrying craft will take off and land conventionally from a standard airport using its jet engines. However, once it is airborne at an altitude of about 12-km, the rocket engines will be ignited to give sufficient acceleration to reach 100-km. In only 80 seconds the craft will have climbed to 60-km altitude. For passenger comfort and safety, it features highly innovative seats that balance themselves to minimize the effects of acceleration and deceleration. The rocket propulsion system is shut down as the ship's inertia carries it to over 100-km, where the passengers will join the very few to experience zero gravity in space.

The pilot will then control the craft using small rocket thrusters to hover weightlessly for 3-minutes and provide passengers with a spectacular view of Earth. After slowing down during descent, the jet engines are restarted for a normal landing at a standard airfield. The entire trip will last approximately an hour and a half.

Astrium is proposing this one-stage system as it is considered the safest and most economical to operate. Development began in 2008, and they say a first commercial flight would be possible by 2012. This project could well be a precursor for rapid transport 'point-to-point' vehicles or quick access to space – opening up previously unexplored territory. Its development will contribute to maintaining (and even enhancing) European competencies in core technologies of space transportation.

As it is a commercial project, private capital will be the main source of funding which could be completed by refundable loans and also by regional development funding. Return on the investment will come from the operations of vehicles for the emerging market of sub-orbital space tourism.

The EADS commitment to offer a plane that will fly at speeds up to and above 3000 miles per hour and travel up to 100-kilometers into space at a cost per passenger of around $200,000 suggests that Virgin Galactic and the other space tourism operators will very likely have stiff competition from Europe. Perhaps more significantly, EADS, as the owner of AirBus, can be expected to be thinking about more than 30-60 minute to trips to nowhere in their business planning. High altitude flights to and from actual cities for corporate executives must clearly be a part of the longer term planning process.[7]

Germany's Project Enterprise

Figure 4.2: Artist's Conception of the Enterprise Space plane in Dark Sky
(Courtesy of the TALIS Institute)

Project Enterprise is a joint venture of the German Space Agency DLR, the German-based TALIS Institute (itself an alliance of five research laboratories and companies), the Swiss Propulsion Laboratory and the multi-national aerospace consulting firm, VEGA. This project is also being carried out with the sponsorship and technical support of ESA. Although there is official participation by the Swiss and German governments and the European Space Agency, the development of the Enterprise space plane is essentially being privately funded. The Enterprise space plane, in theory, could be privately operated. Space flight operations for Project Enterprise are currently planned for a European spaceport. Negotiations have already started on the location, but no final decisions have been taken.

The Enterprise is being designed as a space plane propelled by three modern liquid rocket engines that burn kerosene and liquid oxygen. The craft and its engines will be tested to verify that it is sufficiently safe to take off from a conventional airport against a schedule that anticipates operations by 2010. The European Aviation Safety Authority (EASA) is an active part of this program and will be involved in the testing and independent verification program.

EASA has an established record of promoting common standards of safety as well as promoting environmental protection in civil aviation for Europe. It has spearheaded efforts to create a new, cost-efficient regulatory system in Europe for aviation safety and has played a pro-active role in the project.

The Enterprise space plane is intended to eventually emulate a commercial aviation charter operation. In the same way as the EADS-Astrium project described earlier, the vehicle is intended to take-off like a conventional aircraft and then perform a vertical climb to take it above the atmosphere, reaching a top speed of more than Mach 3. Burn-out of the liquid rocket engines will occur at an altitude of 50 – 80-km. The vehicle will then follow a ballistic ascent path to reach a maximum flight altitude of 100 to 130 km. After burn-out, the passengers will experience weightlessness for a period of 5-8 minutes – depending on the altitude reached. The craft will then re-enter the lower atmosphere for landing back on the original runway. The altitude reached will be sufficient for passengers to see from the North Pole to the Baltic Sea, including all of Central Europe, the United Kingdom, the Alps and the northern Mediterranean Sea.

Earlier studies had envisioned not only carrying space tourists but also scientific missions, but current efforts are exclusively focused on the space tourism market. It is planned to have this spacecraft in service "in three years."

The U.K. – Bristol SpacePlanes

While EADS-Astrium and Virgin Galactic have stolen most of the limelight in Europe, there are also other initiatives which are seeking to be a part of the new space tourism craze and the ambitious Bristol SpacePlanes company should not be overlooked.

The company was created in 1991 by David Ashford and has been involved in space development projects in Britain ever since. Its Ascender vehicle was originally planned to compete for the Ansari X Prize and is designed to carry a crew of two pilots plus two passengers to an altitude of 100-kilometers. It takes off on a conventional airstrip and climbs to altitude of 8-kilometers as a normal jet aircraft and then ignites a rocket engine to climb to an altitude of 64-kilometers and a maximum speed of Mach 2.8. It then coasts on a steep parabolic path to a height of 100-kilometers before descending for a horizontal landing. The steep ascent and descent on the sharply defined parabolic path means that the time of weightlessness will be only two minutes.

The Ascender design is based on an earlier space plane study carried out for the ESA by four U.K. aerospace companies. The design is in many ways a derivative concept from the UX-15 rocket plane and the thrust levels are actually quite comparable. The span of the Ascender is 7.9-meters and the length is 13.7-meters.

The key to the Ascender's operation for a space tourism flight is the near vertical trajectory that achieves the altitude of 100-kilometers, even though the rocket motor thrust terminates at 64-kilometers. The pull out from the returning glide path occurs at an altitude of 46-kilometers

Figure 4.3: Bristol Ascender Space Plane – a UK venture
(Courtesy of Bristol SpacePlanes)

and at its maximum return speed. On the return, the pilot pulls out and gradually flattens the parabolic path to reduce the diving speed using the craft's jet engines. The conventional jet engine then takes over and allows a conventional jet aircraft landing just some 30 minutes after takeoff. The Ascender space plane is thus as much a jet aircraft as it is a space rocket. Bristol SpacePlanes claim that with the Ascender space frame and proven rocket engines, the vehicle would be able to operate several flights a day with a relatively short turn around between flights.[8]

At this time, the U.K. company claims that it will be able to carry out experimental test flights within three years and begin on-going commercial operation of space tourism flights within seven years. They also have a long-range development plan for a new capability called the Spacecab and then moving on ultimately to the Spacebus.

Figure 4.4: The Spacecab space plane departing from the carrier supersonic launcher jet
(Courtesy of Bristol SpacePlanes)

The Spacecab would be an expanded and higher passenger capacity space plane, with two pilots and the ability to carry six passengers or a full crew for the International Space Station. Or it could carry a small satellite for orbital insertion. The Spacecab would be lifted to the upper atmosphere by a supersonic carrier jet designed to fly at Mach 2 speeds. After release by the carrier jet (see Figure}, the Spacecab would reach speeds of Mach 4.

The Spacebus concept is a much larger vehicle, able to carry 50 passengers to a low earth orbit "space hotel." The first stage launcher system would be powered by four high thrust ram-jet engines that resemble the design used in the German Zanger (but avoiding the need for an air-breathing hypersonic scram jet technology.) The launcher, with the docked Spacebus aboard, would climb to 24-kilometers and reach a speed of Mach 2. The Spacebus would then detach and fly into outer space at a speed of Mach 4. The later versions of this craft are envisioned as being able to achieve low earth orbit and then dock with a space station or "space hotel."

Bristol SpacePlanes say that they intend to achieve safety standards similar to commercial airliners. They also estimate that the Spacebus would have operating costs equivalent to about four times that of a Boeing 747. On this basis they project that the cost per flight to orbit and return could be accomplished for as low as $5000 per seat. Although this analysis is highly speculative, even a price of twice this amount would seem to be economically viable, if the independent marketing analysis conducted by the US Futron Corporation (see Chapter 6) is even close to estimating the size of the potential market at variously assumed pricing points. As has been seen time and time again, however, early market projections for totally new services can go seriously wrong.

Projecting demand against future time frames, future prices, and future service options with any degree of accuracy can be dicey at best. Then when one starts to make appropriate adjustments for inflation, the number of competing companies offering space tourism services, and other key variables one can miss the mark between profitability and market failure quite easily. The latest Futron study makes a number of key assumptions that would rather significantly impact any space tourism company's business plan.[9]

Other U.K. Activities
Spacefleet Ltd. is a private company, which was created in 2004 to design and construct vehicles for the space tourism industry. They claim that their current project, a spacecraft called the Spacefleet SF-01, will provide flights in the $120,000 range. The development of the SF-01 is expected to take approximately three years and will cost approximately $260-million for three vehicles.

An interesting feature of their plan is to operate the vehicle in a way that uses electricity generated from sunlight or wind to electrolyze water, in effect producing oxygen and hydrogen. This will then be liquefied using power from the photovoltaic ray. The fuel will then be a non-pollutant and very inexpensive since it is being generated from renewable resources.

The Spacefleet SF-01 will be capable of taking two pilots and eight passengers to heights of 200 miles above the surface of the Earth, nearly three times higher than any other current project in the sub-orbital space tourism industry. It is planned to launch from an inclined ramp and land on a traditional airport runway. It is intended that the spacecraft will be also be usable for quick point-to-point intercontinental travel as well as from a single site.

There were also other projects initiated in the United Kingdom which, like Bristol SpacePlanes Ascender, were started under the impetus of the Ansari X Prize competition and were abandoned after the X-Prize was claimed. A company named Green Arrow planned to have a single-stage manned sub-orbital vehicle; and Reaction Engines

Figure 4.5: the UK's eight seater SF-01 project. Artist's impression
(Courtesy of Spacefleet UK)

Ltd. envisioned using the Skylon air-breathing single-stage-to-orbit space plane in its design. Yet another promising British effort is Manchester-based Starchaser, which planned to develop the Thunderbird manned orbital rocket as a single-stage-to-orbit space plane. Steve Bennett, chief executive of Starchaser, says in his web site:

> "Our overall design philosophy has been to take things one step at a time. We design it, build it, test it and then learn from it. Then we can be sure of the next step."

Finally of course there is the Virgin Galactic enterprise of Sir Richard Branson described earlier in chapter 2. This is a special breed apart. It is neither U.K., US, or Asian in concept and execution. This very high profile space tourism enterprise, funded by Branson's own tremendous wealth, is planetary in its vision. The company's President, Wilt Whitehorn, has indicated that it will likely operate from a number of "home sites." Although nominally a U.K. Company, its SpaceShipTwo craft are being designed and built in the US in southern California. The initial home spaceport for Virgin Galactic will most likely be either the Mojave Air and Space Port, from which Burt Rutan's operations are run, or the new Spaceport America that is under construction as a commercial adjunct to the White Sands Missile Range in New Mexico.

Sweden looks likely to become the first European base for Virgin Galactic's passenger flights (to fly up to see the Aurora Borealis close up and personal). The authorities there say they hope to lower the costs and regulatory barriers to the operation by having it classed as a sounding rocket and given the tax advantages of hot-air balloon flights. At a press briefing in April 2008 at Sweden's Esrange launch site, Sven Grahn, senior advisor to the Swedish Space Corp., which operates the facility in the northern Swedish town of Kiruna, said Esrange's long history as a site for launches of sub-orbital rockets has established a regulatory regime in Sweden to cover third-party liability that also could apply to Virgin Galactic. To reduce the value-added tax that would be levied on Virgin Galactic operations, Grahn said the Swedish Space Corp. is investigating whether the space-tourism activity could be fitted into the same low-tax regime that covers the operations of hot-air balloons.[10]

Virgin Galactic has also indicated that it might consider operations from the U.K., using facilities at existing military bases either in Cornwall in South West England or at Lossiemouth in Scotland.[11]

Russian Space Initiatives – The "Anything Goes" Model

The widely divergent space initiatives in Russia could perhaps be best described as a free-for-all in which everyone is seeking their own part of the prize. A more kindly interpretation is to say that there is now emerging a public-private hybrid model. Russian aerospace companies are working hand-in-glove with the official Russian Federated Space Agency (Roskosmos) and a host of international partners (especially from Europe, but also the US) to develop a wide range of options. Not only are the aerospace companies pursuing these options, but Roskosmos is also willing to be flexible and respond to funding opportunities wherever these might arise. This flexible approach by Russian space officials has been around for some time. For several years, Roskosmos has been saying – over NASA's predictable objections – that they are happy to fly citizen astronauts to space just as long as wannabe space tourists are equally elated to pony up $20/25-million for each ride to space.

Energia and The Klipper

One of the most ambitious initiatives in Russia is that of the Energia Space Missile Corporation. Energia, with several partners, is seeking to develop the reusable Klipper Space Ship that could, in theory, operate to and from the International Space Station by 2015 and perhaps be a part of a lunar transportation system in years to follow. These multi-entry manned spaceships are scheduled to be launched from both Baikonur and the Russian Northern cosmodrome at Plesetsk.

This new space transport program was announced by Energia in April 2006 on the anniversary of the first-ever manned space flight by Yuri Gagarin over 45 years earlier. The Klipper, plus up-rated versions of the Soyuz vehicle, are seen as Russia's answer to low earth orbit service missions for the next 25 years. In time, this new Klipper RLV will replace the well-known and successful Soyuz program. In addition to flights to the ISS, the Klipper is also expected to fulfill other tasks requiring access to earth orbit, providing considerably lower costs for a manned space flight. It is also intended to offer transportation services to domestic and foreign users on a commercial basis.

Figure 4.6: Concept of the Klipper Reusable Vehicle
(Courtesy of Roskosmos)

The Klipper system is to be integrated into the existing space infrastructure that includes the Soyuz-2 and -3 and the Angara rocket system. The modernization of Soyuz spaceships that have been in service for decades is also planned as a part of this program. The existing Progress supply ships will be continued for another five years and are being upgraded with modern, advanced digital equipment. The launch complexes of the modernized Soyuz programs will, in time, also support Klipper and the prospective Angara rockets now under development.[12]

The first phase will be the development of the Soyuz 2-3 program for a new-generation of manned spacecraft. The Samara-based space corporation known as TsKB Progress has announced that their company will build the launcher jointly with Energia. It is currently planned to launch up to an 11-ton spacecraft into 200-km orbits, using the Soyuz 2-3 vehicles. This will be under the auspices of the Federal Space Program that supports both commercial launches, intergovernmental programs and Russian Defense Ministry requirements. According to Roskosmos spokespeople, the rocket's capacity will be upped to 13 tons to launch the Klipper space plane, with a possible further increase to over 16 tons after engine modernization.

A crucial upgrade of the Soyuz launchers will support the Klipper project for a six-person spacecraft to send astronauts into orbit and return them to the Earth on a recurring basis. Upgrades of the system might not only support missions to Earth orbit but in time, potentially, to the Moon or Mars. Separate cargo pods could also be launched on a separate rocket, so that both the Klipper and cargo pods could then be towed to the ISS.

The current schedule calls for the Klipper's first lift-off in 2012, with the first manned flight expected in 2013 and the program becoming fully operational in 2015. This schedule is actually fairly parallel to the NASA Orion project.

International Cooperation with Russia

The 17 member state governments of the European Space Agency (ESA) agreed in mid-year 2006 to participate in a 2-year study project with Roskosmos to explore the design of an Advanced Crew Transportation System (ACTS) that would be capable of missions to the ISS and the Moon. Japan may also eventually contribute to the program, but has not yet signed a binding agreement or dedicated resources to the project. However, JAXA, the Japanese space agency, has indicated that if Europe joined Russia in the program, it would do likewise. At least this is the view of the current status of Japanese thinking, according to Daniel Sacotte, ESA director of the Human Space flight, Microgravity and Exploration program. He said: "There have been no final decisions on the matter, but on the horizon, we think ultimately there will be Europeans, Russians, and Japanese working on this vehicle."[13]

Led by France and Germany, the ESA governments agreed to invest some 15- million euros (about $22-million) for the 2-year effort with Russia to achieve the final design elements and sort out who does what. At this point, it appears that the ACTS will evolve in a two-step process with the Klipper being designed to access the ISS and other earth orbits and then the ACTS will be designed with a capsule return from the Moon. The objective is largely to evolve these programs from existing vehicles to keep development costs to a minimum.

ESA spokespeople have indicated that having two human-rated vehicles that can access the Moon is seen as a key lifeline. This seems quite evident in light of the experience to date with the International Space Station; if there had not been the Soyuz alternative to the Space Shuttle, by this time the ISS would have de-orbited in an uncontrolled and potentially disastrous manner.[14,15]

Myasishschev Design Bureau

Russia's Myasishchev Design Bureau (MDB) has a long history of aerospace development for both advanced aircraft and space vehicles. In particular, it is famous for its strategic bombers the 3M, the M-4, and the M-50. It is also responsible for the development of the cabin for the Buran orbiter, for flight testing during the development of VM-T Atlant transport aircraft, for its high altitude aircraft M-17 Stratosphera and for the M55 Geophysica.

In the space tourism area, MDB is currently involved with upgrading the Cosmopolitan XXI high altitude jet into an operational Explorer Space Plane (the C-21) in partnership with the US-based Space Adventures company. The C-21 will also operate from the MX-55 High Altitude launcher plane to create a horizontal take-off and vertical parachute landing (HTVL) space plane system. MDB is also developing the same capability, namely the C-21, with the MX 55 high altitude launcher plane for another US client, the Sub-Orbital Corporation. The second stage will be the C-21, a rocket-powered lifting body with parachute landing.

Figure 4.7: The C21 joint venture spaceplane from MDB and Space Adventures
(Photo courtesy of Space Adventures).

Space Adventures pioneered the space tourism business by teaming up with Russian partners, and especially Roskosmos, to delivery citizen astronauts to orbit. But Eric Anderson, President of Space Adventures and ultimate space entrepreneur, has also teamed with XCOR to seek another option in terms of developing its Xerus space plane as an alternative sub-orbital launch capability. Eric has time and time again shown his prowess in developing not only a Plan A but also a Plan B for his company's strategic planning process.

The Chinese Space Program

China reached a milestone in space history in October 2003 with the launch of its first piloted space flight into Earth orbit. Blasting off from the Jiuquan Space Launch Center in Inner Mongolia atop a Long March 2F rocket, a Chinese astronaut named Yang Liwei went into orbit around the Earth aboard the Shenzhou 5 spacecraft, according to the official Xinhua News Agency.

Chinese president Hu Jintao was at the launch site to witness the launch in person and said: "The party and the people will never forget those who have set up the outstanding merit in the space industry for the motherland, the people and the nation."

As a result, China became only the third nation on Earth capable of independently launching its citizens into orbit. The former Soviet Union was first in 1961, followed by the United States in 1962. The space capsule, whose more modern design is largely based on the Russian Soyuz spacecraft, made 14 orbits and remained in space for about 21 hours before executing re-entry and a parachute landing onto Chinese soil.

Experts commented that China's space infrastructure, its array of launchers, its space industry and now a piloted space mission placed them above even the Japanese, in terms of demonstrated space capabilities and put them in the same category as the USA and Russia.[16]

Another landmark was achieved at the end of September 2008 when another astronaut, Zhai Zhigang, became the first Chinese person to walk in space, during the his country's third manned mission. In live satellite television pictures broadcast in China and around the world, Zhai – the 41-year-old son of a snack seller chosen for the first "extra-vehicular activity" – waved a small Chinese flag, helped by colleague Liu Boming who also briefly popped his head out of the capsule.

Zhai returned inside the craft safely after about 15 minutes. He wore a $4.4m Chinese-made suit weighing 120kg (265lb). Liu wore a Russian-made suit and acted as a back-up. The manoeuvre is regarded as a step towards China's long-term goal of assembling a space laboratory and station.

The Shenzhou VII craft blasted off from a remote site in the Gobi desert in the north-west of the country and landed safely in Inner Mongolia. Afterwards, China's Communist party leaders revelled in the positive publicity the space mission received, particularly after Beijing's successful Olympics. "On this flight, Chinese people's footprints will be left in space for the first time," said a commentary by the official Xinhua news agency. "This will give the world yet something else to marvel about China in this extraordinary year of 2008."

China's economic, political, and technological programs have rocketed forward in figurative and literal ways over the past decade. When the author was Dean of the International Space University attending the International Astronautical Federation meeting in Beijing, Vice Minister Li Pen gave a briefing to the assembled heads of various space agencies from around the world. He outlined current and planned space initiatives that China would undertake in the next two decades. He talked about communications and other application satellites, new and improved launchers and even trips to the Moon and beyond. He spoke in great depth about all of these programs and answered questions—all without any notes and with ready knowledge about both space applications and manned space programs. This was not a high level political leader reading from a script, but a person dedicated to a mission.

China has been clearly dedicated to becoming a space power of the first rank for decades and the results of this effort are now apparent. Since that time China has, if anything, redoubled its efforts. Its astronauts have flown safely into earth orbit. A Chinese engineered and manufactured communications satellite has been launched to meet the domestic telecommunications needs of Nigeria. Cooperative projects have been undertaken with Brazil. Numerous applications and scientific satellites have been launched to meet localized needs in meteorology, TV, communications, remote sensing and more. A robotically-deployed project to build an observatory on the Moon in cooperation with Italy and other European partners has been announced and lunar exploration by Chinese astronauts is no longer a vision but a serious program. The Chinese test of its anti-satellite (ASAT) capability by targeting and destroying an obsolete meteorological satellite has most recently captured global attention.

The China Academy of Launch Vehicle Technology (CALT) is now engaged in research to develop a first-generation reusable launch vehicle (RLV). The institution, which develops launchers for China's space program, unveiled its road map for RLVs at the International Astronautical Congress in Valencia, Spain in September 2006. This long term, 20-year, Chinese launcher development plan envisages three generations of RLVs. The first is a partially reusable two-stage-to-orbit (TSTO) launcher. This is to be followed by a fully reusable vehicle with vertical take-off and horizontal landing for the second stage. The third launcher in this longer-term plan is for a fully reusable aircraft-like single-stage-to-orbit (SSTO) vehicle with horizontal take-off and landing. CALT research and development centre manager Yong Yang, in his presentation at the International Astronautical Conference stated: "We don't have an official schedule for the programme, but the first generation could be developed in 15 years."[17]

CALT expects to conduct a Mach 15 re-entry demonstration flight using a winged upper stage that will be launched to an altitude of 100 km (62 miles). Another planned demonstration is an automatic flight and landing test to prove guidance, navigation and control systems. No dates have been provided for these tests. In the short to medium term, meanwhile, China's strategy for launcher development involves cryogenic liquid oxygen / liquid hydrogen expendable launchers using a common booster architecture. The cryogenic propellants will replace the solid and other toxic fuels used in the country's existing expendable rockets.

In early 2007, China's State Council approved the country's 11th five-year plan on space development covering the period from 2006-2010 with manned space flight, lunar exploration, new launch vehicles and high-resolution earth observation being areas of priority. China is expected to develop nearly 100 spacecraft during this period. This could well make China the world leader in overall space activities if the full five-year plan is achieved.

Their focus in space is on three primary areas where use of the medium will give Beijing an overwhelming advantage; these are satellite navigation systems, remote sensing and space communications. The interlinking of these technologies is designed to achieve what the Chinese are now calling a spatial information super highway. This will include a network of communication and broadcasting, earth resource, meteorological, navigation, scientific experiment satellites, and so on. China hopes to involve the private sector to sell these technologies in a big way and develop the space service sector.[18]

Not too surprisingly, the officially sanctioned Chinese space development program is oriented toward development capabilities to support governmental programs, rather than to pursue commercial space tourism programs. But as the space tourism business develops, it is possible that some of the Chinese CZ and Long March vehicles might be adapted to these purposes, but there are no announced initiatives in these areas at this time.

Canada's Commercial Space Program

If there is a clear counterpart to the US space commercialization endeavors, but on a smaller scale, Canada probably best fills the bill. For a country of only 25- million or so residents, Canada is churning with space plane initiatives.

The Canadian Arrow Corporation is a privately-funded rocket and space travel project founded by entrepreneurs Geoff Sheerin, Dan McKibbon and Chris Corke. Based in London, Ontario, they had been considered one of the top three candidates for the X-Prize competition before losing out to Burt Rutan in October, 2004. Their two-stage vehicle was a 16.5-meter (54-ft) long three person sub-orbital rocket with the second stage doubling as an escape system. It was designed with four main parachutes to slow the second stage for a splashdown on water at approximately 30 ft/sec.

Canadian Arrow received enough encouragement from the Canadian Government and private investors to continue their quest to be a leader in the new space tourism business. During 2006, the company joined forces with Planetspace based in Chicago and recruited a team of test astronauts for their spacecraft program. In August 2006, PlanetSpace / Canadian Arrow announced an agreement with the province of Nova Scotia for a 300-acre launch site and training center at Cape Breton. This agreement included financial incentives from the Nova Scotia government and will involve the transition of their launch vehicle testing operation from the Meaford military installation in Ontario to Cape Breton.[19]

PlanetSpace, headed by Dr. Chirinjeev Kathuria, is part of the new emerging space commercialization business. It has also teamed with major players in rocket booster and space craft development as it seeks to develop a broad spectrum of commercial space services that include cargo and crew to the ISS, point-to-point global travel, space tourism and satellite orbital delivery. Lockheed Martin Space Systems Company, is developing the Orbital Transfer Vehicles for PlanetSpace; and ATK Launch Systems, leads the development of the launch vehicle segment and ground processing systems.

Planetspace is also developing the Silver Dart hypersonic glider designed to operate as an orbital vehicle. This is a two-stage spacecraft, using the Nova launch system. The Silver Dart is designed to double as an unmanned or manned spacecraft and can provide for a long duration platform in orbit. Based on the FDL-7 design which

Figure 4.8: Canadian Arrow Vehicle
(Photo Courtesy of Planetspace)

Figure 4.9: The Silver Dart design on display
(Photo courtesy of Planetspace)

is stable in flight from Mach 22 to 0, it has the glide range of 25,000 miles (one Earth circumference). An all-metal thermal protection system allows for all weather flying. Combined, the thermal protection system and glide range result in a reentry vehicle that cannot be trapped in space and is able to return to base from any orbit around the Earth.

The PlanetSpace / Canadian Arrow web site says it plans to provide sub-orbital flights for 2,000 citizen astronauts within the first five years of operations, offering not only sub-orbital flights but, it hopes, to flights to the International Space Station as well. Plans for the new Cape Breton site call for a massive orbital launch facility that will involve industry giants and could eventually be on a scale comparable to NASA's operations at Cape Kennedy. "We're basically building a private manned space program for Canada," says Dr. Kathuria.[20]

Dr. Kathuria is no stranger to privately funded space flight. He was a founding director of MirCorp, the Russian company that made history on April 4th, 2000 when it launched the world's first privately funded manned space program and signed up Dennis Tito as the first space tourist (or citizen astronaut). MirCorp was a joint venture with RSC Energia, which launched the first satellite (Sputnik), sent the first man to orbit the Earth (Yuri Gagarin), built the Mir Space Station, and is a major partner in the ISS.

The company says they expect to generate annual revenues of $200-million (US) in their fifth year of operations. Fares will start at $250,000 for a sub-orbital flight, including fourteen days training. According to Mr. Sheerin:

> "Eventually, our goal is to make Planetspace a public company. Space should be open to all people, and all people should also have the opportunity to invest in the future of space travel."[21]

The Da Vinci Project

Another of the new initiatives in Canada that was born of the 2004 X-Prize competition was the so-called da Vinci Project. This project envisioned launching a small 18-meter long craft to a height of some 60,000-feet via a balloon. The craft would then disconnect from the balloon and rocket to 100-kilometers.

In 2006, Brian Feeney, the Canadian founder of the da Vinci Project, launched DreamSpace™, a new brand aimed at the space tourism market. The objective is to create a line of successively larger spacecraft, capable of providing personal space flights to an ever-increasing tourism base.

The technical team behind DreamSpace™ comes from the original da Vinci Project's team of aerospace engineers and project personnel. This team introduced the XFI as the first of a planned line of multiple passenger spacecraft. The one-person technology demonstrator will be used to constantly broaden the space flight envelope, in preparation for the DreamSpace™ goal of space tourism. Technical details for this new ship were released at the 2006 X-Prize Cup event at Las Cruces, NM, where Brian Feeney said they were aiming for test flights in 2007.

Figure 4.10: The Da Vinci Rocket
Marketed by Dreamspace
(Courtesy of the da Vinci Project)

The XFI employs a new liquid rocket propulsion system and aircraft turbine engine. The objective of the Da Vinci Project is to develop a winged space aircraft design able to operate out of an airport-style spaceport with airline efficiency. Such objectives and visionary aims are easy to put on a web site or proclaim at a news conference but hard to deliver in practice.

Australian Projects

Australia was involved in early space research and rocket development. It was indeed the third country to have a satellite launched – from its Woomera Range in the 1960s. However, it abandoned space activities as a national government enterprise some years ago on the basis that space capability is "not essential" to its strategic longer-term mission. The current Australian national strategy is to encourage private enterprise to develop launch capabilities instead of developing a national rocket program or governmental operated launch facility.

This policy decision has led to a variety of different activities undertaken by private institutes, universities, and research organizations. These initiatives include those of the Australian Space Research Institute (ASRI), the Australian Rocket Program (AusRoc), the Cooperative Research Centre for Satellite Systems and the Queensland University of Technology. The AusRoc program, sponsored by ASRI and corporate and private donations, has developed the AusRoc 1, 2 and 2.5 sounding rocket programs. Efforts are also currently underway to develop the AusRoc 3 and 4 programs. The AusRoc 4 is intended to launch a small satellite of 35 kg into low earth orbit.

Virtually all of these launcher developments are centered on the launch facility in Woomera and it appears likely that a new Australian enterprise will evolve out of these activities. But at this time the most significant efforts to upgrade the launch center come from US initiatives. At this time, some $100-million (Australian) is being spent on the Woomera launch site for its modernization and upgrade.

Both Rocketplane-Kistler and DTI Associates have signed agreements to use the Woomera launch facility. At this stage, DTI Associates is proposing to use the Terrier-Orion vehicle to launch cargo (up to 300 Kg) into low earth orbit, while RpK had planned to use the facility for manned launches. As described in chapter 3, RpK was supported by NASA's COTS R&D program for payloads to be launched to the International Space Station, until this was terminated by NASA in late 2007. Their K-1 vehicle was planned carry its payload into orbit, after which the upper stage would return to Earth near the launch site. The second stage would return to Woomera about 24 hours later.

Indian Initiatives

The Indian Space Research Organization (ISRO) and the Indian Defense Research and Development Organization are currently developing a reusable launch vehicle program (RLV). It is their objective to develop such a RLV capability in order that launch costs can be significantly reduced.

The project is underway at the Vikram Sarabhai Space Center to develop an alternative lift capability to the current fleet of Indian expendable launch vehicles. According to space scientists, their reusable vehicle would have a three-fold advantage. These would be to allow lower cost launches, more frequent missions and more international customers for launch services. Currently, ISRO's cost to place a kilogram of payload in orbit ranges between $12,000 and $15,000. Projected costs for the new launch capability might reduce these costs by as much as ten times to the level of $1,200 to $1,500. The first technology demonstrator flight of the new reusable Indian vehicle is in the 2008-09 time frame.

Studies for the propulsion system for the new launcher have focused on a semi-cryogenic engine with a two-stage-to-orbit (TSTO) vehicle. The design envisages the first stage being configured as a winged body system that will attain an altitude of 100-kilometers. At a particular point it will burn out, re-enter the earth's atmosphere and land horizontally on a runway like an aircraft. After delivering the payload, the second stage will re-enter the atmosphere and land. This new reusable vehicle has been called the Indian version of the space shuttle. Indian space officials say that their reusable vehicle will be superior to the shuttle because it will have a number of enhanced safety features and be much more cost effective. The new vehicle is being designed primarily to lift satellites into orbit, but it could evolve into a craft to carry human passengers with a splash down capsule.

This program will clearly be a number of years in development and will not become operational until the post 2012 time period. It would not be intended to launch a crew until the much further into the future. At this time, the program is entirely governmental in nature without commercial involvement. The first stage of the vehicle might be sold to commercial interests just seeking a sub-orbital flight to an altitude of 100-kilometers.[22]

The Israeli Space Plane

Even Israel has attempted to be a part of the new space tourism industry as the fever to go into space has spread worldwide. The Israeli Negev 5 system to provide access to sub-orbital space is the design of Dov Chartarifsky of IL Aerospace Technologies (ILAT). It is an ingenious combination of a hot air balloon and a solid rocket motor – an unconventional concept employing proven and conventional technology to accomplish the mission.

The Negev-5 is a hybrid "spacecraft," to be launched from the ground using a hot-air balloon as a first stage. After being dropped from the balloon at an altitude of 10-km (6.25-miles), the spacecraft would fire a solid rocket motor for 96 seconds. At engine burnout, the vehicle would be traveling at Mach 3.4 and the crew would experience weightlessness for four minutes, reaching a maximum altitude of 120 km. The passengers would experience four minutes of weightlessness before descending to an ocean landing by parachute. At an altitude of 5-km, drogue and main chutes would deploy sequentially and the main chute would slow the capsule down to 7 meters/second (i.e. about 24 ft/second) before splashdown.

Republic of Korea

The Republic of Korea has been increasingly a player in the aerospace field over the past two decades. South Korea has made a solid move into the world market for automobiles and has essentially said if we can do cars, then why not rockets too. Samsung, LG Corporation and Hyundai have all developed space capabilities as well as governmental research labs. Projects have ranged from the Hyundai Corporation serving as system integrator (i.e. for the Globalstar mobile satellite system) down to microsatellites undertaken by South Korean universities.

The major initiative in the spaceplane field currently underway comes from a the Korean-based rocket company called Challenge & Space Industries (CSI) that is teamed with AirBoss Aerospace Inc. (AAI). This joint US / Korean effort, based on a Memorandum of Understanding (MOU) signed on August 17, 2005, is seeking to develop a sub-orbital space plane called Proteus. The Proteus is a Vertical Takeoff and Horizontal Landing (VTHL) vehicle powered by turbo-pumped liquid oxygen and methane fuel engine, known as the Chase 10 and developed by CSI. This regeneratively cooled engine can produce 22,000 lbs of thrust and has demonstrated reliability in extended tests. Further details and pictures can be found in *Aviation Week and Space Technology* magazine (www.aviationweek.com).

AirBoss, the US partner, is serving as the system integrator and is also developing the airframe that will be used in this Delta-winged space plane. The initial version of the craft has been designed to carry a crew of three, but its progress has been delayed by US ITAR regulations concerning the transfer of sensitive technology. Ion Aircraft and Phoenix Technology Works are also a part of this development initiative. (There is also speculation that in the 2010 to 2015 time frame, C&Space will also be involved in the design of a new Korean KFX "stealth" fighter that will also use their liquid oxygen and methane engineer.) Clearly in many international space plane developments, including those in the Republic of Korea, there is an interest that goes beyond space tourism and spills over into military / defense interests as well as supersonic commercial transport.[23,24]

Argentina: Why Can't We Also Get Into the Space Tourism Business?

Argentina's so-called Gauchito "Little Cowboy" project sought to develop a manned sub-orbital flight vehicle. This project, however, was discontinued in 2004. It was originally undertaken in response to the X-Prize contest, but was unsuccessfully tested at the half-scale model size. The vehicle was designed to use a hybrid rocket with a powered vertical take-off and parachute descent to a water landing. It was conceived by Pablo De Leon and Associates of Buenos Aires.

Pablo de León is a man of some vision. He is the co-founder and president of the Argentine Association for Space Technology (AATE), and a graduate of the International Space University.

The Gauchito was envisioned as a clustered design made up of four hybrid engines producing a maximum acceleration of 3.5G. The engines would burn until the vehicle reached an altitude of 34-km, followed by a coast to 120-km. After five minutes of weightlessness, the pilot was to orient the thermal shield for re-entry. De Leon and Associates flew a subscale test capsule to 30-km and conducted one-third scale rocket test firings but a half-scale rocket test firing resulted in a failure. Testing of flight suits and other critical systems were also conducted but this program was discontinued after the Ansari X-Prize was successfully claimed in 2004.

International Space Hotel Initiatives

Most international efforts have been aimed at developing spaceports or space vehicles to support the space tourism industry. But most recently Galactic Suites of Barcelona, Spain, has announced the incredibly ambitious $3-billion initiative to deploy a space hotel. This initiative has backing from Spain, the US and other countries and aims to deploy such a facility by 2012 and charge $4-million for a space stay. This project has a long way to go to catch up with the Bigelow initiative, but the approach they are taking to use a cluster of "cells" for space tourists is clearly much different than the Transhab technology.

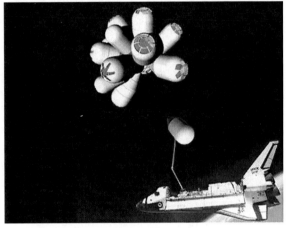

Figure 4.11: Artist rendition of the Galactic Suite "Space Hotel" in Orbit
(Courtesy of Galactic Suite Co.)

Conclusions

The international dimensions of the space tourism business can only continue to grow. The demand for space travel is certainly global. The market is not restricted to any nationality, creed or belief system. Spaceports may dominate the first wave of globalism, but as can be seen by the many projects described above, efforts to develop space plane technology and to undertake needed research will continue to expand around the world as well. There are several recurring themes that can be seen in this global review. These include the various ways that private enterprise and governments cooperate, collaborate and even compete – all at the same time. There are also areas of strategic concern that involve possible defense or military use of space plane technology as well as issues relating to intellectual property and the US ITAR regulations. The most likely space tourism ventures to succeed may very well be those that can raise capital from a global market and tap into a wide range of scientific and engineering knowledge from around the world. Virgin Galactic, Galactic Suites and Space Adventures are clearly designed as multi-national ventures that are following such a multi-national model.

Finally there are unresolved issues about how space plane flights will be regulated as well as who will control and oversee the rising number of spaceports. There appear to be at least some unresolved international and national legislative and policy issues, despite actions undertaken by the US Congress and the FAA to establish an initial framework for such flights in America. Prime among these issues is whether the International Civil Aviation Organization (ICAO) or some other body is in charge of international safety standards?

Just as the diversity of technical approaches creates a global laboratory against which one might test the best space plane technology, the international community appears to be beginning to form a sort of "international policy framework" against which space tourism safety and regulatory controls will be tested as well. With literally thousands of citizen astronauts on the verge of flying at least into sub-orbital space within the next few years, some of these safety and regulatory control issues will need to be answered sooner rather than later.

Further, the entry of EADS Astrium, the owner of AirBus, into the high altitude transport business suggests that space tourism may give birth to high altitude, supersonic air transport for corporate executives and high rolling jet setters. This has several implications. The most serious of these may be the air pollution factor for the stratosphere and how this new industry may give rise to issues that go beyond global warming. Destruction of the ozone layer might prove to be even deadlier to humans than the rise of carbon pollutants. This is an international issue deserving paramount attention. It is ironic that Sir Richard Branson who has offered top prize money to address global warming issues, may be the father of a new industry that could be a source of deadly pollution. Go Figure.

References:
1. ESA – General Studies Programme – www.privatespaceflight@esa.int
2. Orbit report – http://www.orbireport.com/Linx/Startup.html
3. The Guardian, U.K. – February 7, 2008.
4. Phoenix Reusable Launch Vehicle – http://www.space.eads.net/families/access-to-space/launch_systems/future-launchers/view?searchterm=Phoenix
5. Abul Taher, "Europe Joins Space Tourism Race," Times of London, June 10, 2007
6. EADS/Astrium press release – Wednesday, June 13, 2007
7. Abul Taher, "Europe Joins Space Tourism Race," Times of London, June 10, 2007
8. Bristol Spaceplanes Ltd. – http://www.bristolspaceplanes.com/projects/ascnder.shtml
9. Bristol SpacePlanes Ltd. – http://www.bristolspace plane.com/projects/spacebus.shtml
10. SpaceNews.com, April 1, 2008
11. Carl Hoffman, "Now Boarding" – Wired, June 2007, p. 152-153.
12. Klipper Program Announced, ITAR-TASS, April 11, 2006
13. The Planetary Society -
    http://www.planetary.org/news/2006/0628_Europe_and_Russia_Join_Forces_to_Study.html
14. The Planetary Society -
    http://www.planetary.org/news/2006/0628_Europe_and_Russia_Join_Forces_to_Study.html
15. Astro News – "Development of New Soyuz 2 and Soyuz 3 Rockets" May 12, 2006
    http://www.astroexpo.com/news/newsdetail.asp?ID=25688&
    ListType=TopNews&StartDate=5/15/2006&EndDate=5/19/2006
16. Space.com – October 15, 2003 – http://www.space.com/missionlaunches/shenzhou5_launch_031014.html
17. Rob Coppinger, "First RLV by 2020?" – Flight International, October 17, 2006,
    http://www.flightglobal.com/articles/2006/10/17/210002/first-rlv-by-2020.html
18. Rahul K Bhonsle – 2007 NewsBlaze, Daily News
19. "Cape Breton joins space race: N.S. signs private launch facility deal" – Toronto Star, Aug. 16, 2006.
20. Planetspace web site – www.planetspace.com
21. Planetspace web site – http://www.planetspace.org/lo/index.htm
22. Srinivas Laxman, Reusable vehicle to cut costs, attract business, Times of India, January 11, 2007 –
    http://timesofindia.indiatimes.com/Reusable_vehicle_to_cut_costs_attract_business_for_India/articleshow/1129051.cms
23. Hobbyspace.com – http://www.hobbyspace.com/nucleus/index.php?itemid=207
24. AsiaFinest.com – http://www.asiafinest.com/forum/index.php?showtopic=76498

— Chapter 5 —

"Spaceports are popping up everywhere, but they have nothing to fly ... and with investments spurred by very little information. It's almost humorous to watch the worldwide battle of the spaceports ... they're everywhere." – Burt Rutan[1]

THE SPACEPORT STAMPEDE

The prospect of private space flight has led to almost a glut of "so-called" spaceports in the United States and around the world. Under US jurisdiction alone there are seven Federally Operated Spaceport Facilities – the Cape Canaveral Air Force Station, Edwards Air Force Base, NASA's Kennedy Space Center, the Kwajalein Island tracking and telemetry facilities, Vandenburg Air Force Base, the Wallops Island, Virginia facility and the White Sands Missile Range.

In addition there are, at this time, six fully Federally licensed spaceport facilities that all have considerable capability and are available to support private space vehicle programs. These include the California Spaceport, the Kodiak, Alaska Launch Complex, the Mid-Atlantic Regional Spaceport (MARS), the Spaceport operated by the Florida Space Authority at Cape Canaveral and the Southwest Regional Spaceport Project located in Upham, New Mexico.

Then, add to these the growing number of regional facilities in the US that are vying to provide spaceport operations in support of new private space launch operations. These include Brazoria County, Texas, the Clinton-Sherman Air Force Base in Oklahoma, the Willacy County Texas facility, Spaceport Alabama in Baldwin County, Alabama, Spaceport Washington in Grant County, Washington, the West Texas Spaceport in Pecos County, Texas and the Wisconsin Spaceport in Sheboygan, Wisconsin.

Figure 5.1: Artist's Conception of Citizen Astronauts – waiting to fly into space
(Artist's impression – courtesy of Futron).

Then beyond the US, there are numerous launch facilities operated by the European Space Agency and CNES of France. There are three state-run facilities in China, as detailed later in this chapter, and other launch sites operated by the national space agencies of the Russian Federation, the Ukraine, Japan, and India. And then there are new commercial spaceport projects, which have been announced by several other countries including Australia, Canada, Israel, Malaysia, Singapore, the U.K. and the United Arab Emirates.

But even in the unlikely event that *all* the planned and potential new space tourism operators are successful in meeting their projected timetables and passenger forecasts, it is difficult to see how most of these new facilities could operate profitably. The thought process of many of those developing, promoting or backing these projects must be that of "loss leaders." These backers seem to be thinking that there is money to be made in other ways. For example, some states in the US see a spaceport as a job magnet for high paying "green jobs" that will come as aerospace, computer and other high tech companies flock to be associated with this exotic new business. State economic development units are eager to not be left behind and present themselves as forward thinkers and "high flyers" in the new frontier of space. These states such as Alaska, California, Florida, New Mexico, Oklahoma, Texas, Virginia and Wisconsin have not only supported to creation of one or more spaceports but in several cases have enacted new legislation to bring state rules into line with federal laws concerning liability claims and public safety rules.

The Theme Park Model

Others have a more elaborate concept of what a space port complex is all about. Space Adventures, the leader in space tourism to date, has focused its planning on four sites – in Florida, Las Vegas, the United Arab Emirates and Singapore. The US sites are clearly top tourist destinations, already associated with places where vacationers go not only to relax, but also to be entertained, let off steam, and experience something new and different. At a typical theme park a family of four will spend hundreds of dollars within the amusement facility. In terms of a business model it is probably important that at the same time they will probably spend thousands of dollars at local hotels, restaurants and retail shops.

As an "entertainment enterprise" the spaceport business case becomes ever more complex and challenging. Hollywood has actually had to re-invent its business model several times over in the past few decades. Movie studios such as Universal make less money on blockbuster movies like Jurassic Park and Lost World than on the associated rides at their amusement park.

Consider the basic economics. Someone going to a movie spends about $10 or only $4 if it is a rental. Someone buying a "Lost World" computer game might spend $60. But someone going to an amusement park, staying at a hotel and eating at restaurant, paying for parking, and shopping for a new swimsuit and souvenirs may easily drop a thousand or more. You do the math. The movie studios already have. The successful "space park / space port" operators will try to do the same – that is turn the space ride experience into a reliable and large revenue stream.

Eric Anderson of Space Adventures, no dummy when it comes to inventing new entrepreneurial business models, has already given this issue some serious thought. He feels that he might be able to franchise high tech spaceports that are a new type of 21st century amusement park. The two facilities already under serious planning in Singapore and the United Arab Emirates, where potential partners have some very deep pockets, will look and feel like a totally new animal – part air and space port, part Disney World Epcot Center, part game arcade, part hotel, part restaurant, and part shopping mall.

These new spaceports will have lots of interesting features. The elaborate and capital intensive facilities could easily include space simulation facilities, space training and fitness centers, as well as take off and landing terminals for space flights. But this is just to get started.

Closely associated with the spaceport might well be a space amusement park with rides and experiences focused on the future and space, with lots of corporate tie-ins and product placement opportunities. Although the high-end space tourists are expected to pay $200,000 to $250,000 per ride and then spend perhaps another $50,000 in "extras," a much larger "gate" of family tourists and fun seeking vacationers could be attracted to a glittering spaceport for some low altitude rest and relaxation and a few thrills that will cost a lot less. What are these possibilities?

Possible options are already percolating through the fertile minds of X-Prize executives such as Peter Diamandis and Gregg Maryniak. You can currently take one of Peter's Zero-G flights for under $4,000. This high arc in the sky can let fun seekers experience about 40 seconds of weightlessness. This can be extended to minutes of floating in space if you sign up for multiple parabolic arcs, such as Professor Stephen Hawking ultimately took. Zero-G is closely aligned with Space Adventures and operates from both Florida and Las Vegas. But there is far more than parabolic rides in a jet.

There are the X-Prize space demonstrations and potential "space races" such as those already planned to be staged in Las Cruces, New Mexico. The NASA Centennial prize competitions are already part of the "rocket to the future" fun.

Meanwhile the FAA regulators are having fits trying to figure out the safety controls that need to apply to all of these activities. The regulators are trying to see how one can best combine serious, high explosive rocket systems on one hand with tourism, entertainment and performances geared to mass market audiences on the other. NASCAR races bring excitement and close up views of high octane race cars locked in dangerous combat. Protective barriers and other precautions offer reasonable protection to the grandstand audiences. Pit crews, race officials, firemen and other racetrack workers have exposure to crashes and other risks; but fatalities, other than among the race drivers themselves, have been few and far between. The FAA officials are not sure whether NASA and Air Force guidelines for launch operations are the best point of departure for "Space Races and Shows" or whether NASCAR races should be the starting point for regulating safety. To date the FAA has been trying to set safety guidelines that leverage off those for air shows, where safety standards have not necessarily been all that successful.

This much is clear. A number of "spaceport developers" are seriously planning to expand the facilities into space-related leisure and recreation centers along the lines of Epcot or even Disney World. The list of possible offerings only continues to expand:

- Air and Space Museums.
- Elaborate educational and scientific exhibits.
- Thrill-a-minute simulated space flights for the public.
- Space X, NASA and other "challenge prize competitions" to test or prove the capability of new state-of-the-art lunar and Mars landers, new rocket planes and scram jets demonstrations.
- Parabolic flights to achieve weightlessness.
- Rides in extremely high altitude jets.
- High G centripetal force centrifuge rides.
- And, potentially, space rocket races within "marked courses" in the sky.

Space ports or Recycled Airports

As already mentioned, a number of US states, bent on economic development and seeking a new high-tech cachet, have latched on to what they see as a very appealing idea. Their plan is to spruce up discontinued airfields into shiny new facilities. They hope to turn their down-at-the-heel airport Cinderellas into shining spaceport princesses.

Already there are a burgeoning – and perhaps disturbing–number of decommissioned US Air Force bases and shut down airports that are being re-invented as spaceport facilities. This, at first sight, seems a great idea. It provides a new use for abandoned or under-utilized locations. Further, these locations already have a great deal of expensive infrastructure in place and it gives a high tech flair to proclaim it as a spaceport. Advocates see the opportunity to attract cutting edge industry, new jobs, and an infusion of scientific talent.

There is unfortunately a flaw in the thinking. This is simply that there are only a very limited number of space tourism companies that will succeed. The proliferation of spaceports will soon outnumber the potentially viable space plane ventures, perhaps five to one. As Burt Rutan has bluntly observed, a spaceport without any flights is probably going to fail.

The transformation of old airports into spaceports is one thing, but making them successful enterprises is another. Renovating these facilities and transforming them into high-tech, exciting "amusement parks" is a gigantic undertaking. Just to calibrate such an effort, the Walt Disney Corporation's creation of a new park outside Tokyo – now almost a decade ago – involved a capital investment of $4-billion. The creation of a spaceport and a "space park" is an effort that takes time, requires environmental impact statements, extensive planning and engineering and potentially billions of dollars of investment over a period of many years. Such an effort involves not only planning, building and staffing the "space amusement park facilities" but also the creation of the rest of the infrastructure from hotels and restaurants to water and sewer systems and so on.

The FAA Regulation and Licensing of Space Ports

Since 1996, FAA / AST has licensed the operations of six non-federal launch sites and now many more are in the pipeline. These new spaceports are expected to serve both commercial and, potentially, government payload operators. To date, according to the FAA / AST 2007 Report 'Commercial Space Transportation Developments and Concepts, Vehicles, Technologies, and Spaceports', about $200-million has been invested into non-federal spaceports across the US This figure is probably conservative. It does not include the $100-million (Australian) that has been invested in the Woomera launch facility and spaceport that has been upgraded and modernized to accommodate the US-based Rocketplane-Kistler and DTI enterprises. Nor does it take into account the new authorizations approved in 2006 by the States of New Mexico, Florida, Virginia, Maryland, etc. for upgrades to commercial spaceports.

In short, capitalization associated with US and overseas sites continues to grow rapidly and may double in the next year or two in response to new initiatives. In the US, current activity to establish or upgrade spaceports is primarily funded by the individual states, augmented by private sponsorship and some federal government support. It is estimated by the FAA / AST that an average of about $3-million is being spent yearly at each facility to operate these established and licensed spaceports. Again these operating cost estimates appear to be conservative.[2]

The table below shows the eleven US states that have federal, non-federal, and proposed spaceports. While the majority of licensed launch activity still occurs at US federal ranges, much future launch and landing activity may originate from private or state-operated spaceports. For a non-federal entity to operate a launch or landing site in the US, it is necessary to obtain a license from the federal government through FAA / AST. Of the six non-federal launch sites licensed by FAA / AST, three are co-located in conjunction with federal launch sites. These are the California Spaceport at Vandenberg AFB, the Florida Space Authority spaceport at Cape Canaveral and the Mid-Atlantic Regional (MARS) Spaceport (originally the called the Virginia Space Flight Center) at the Wallops Flight Facility, Virginia.

In the United States alone, political and financial muscle is at work to install new spaceports in a number of states as listed earlier. Texas, in its tradition of being the biggest in all its ventures, now has three spaceport projects in various stages of planning and licensing. These states and a growing number of others are busy pushing tax credits, free land and facilities and other incentives to get developers to build a shiny new spaceport in their backyard. This proliferation of spaceports can only lead to a number of stresses and strains, if not serious problems. One obvious observation is that this number of spaceports is going to stress the capabilities of the FAA-AST staff to complete an in-depth inspection of all the facilities before they can be licensed.

State of Operation	Federal	Non-Federal Facility (Licensed)	Proposed
Alabama			X
Alaska		X	
California	X	X	
Florida	X	X	
Kwajalein (Marshall Islands)	X		
New Mexico	X	X	
Oklahoma			X
Texas			X, X, X
Virginia	X	X	
Washington			X
Wisconsin			X

Figure 5.2: Status of Spaceports Operational or Proposed in the US

Meanwhile, on the world scene, in addition to the existing space agency launch sites in Russia, China, India and French Guiana, there are the other commercial or hybrid projects mentioned earlier. There is the existing Woomera site in Australia and in Scotland a spaceport has been touted for a former military airfield. There are also advanced plans to build spaceports in Canada, and the two projects in the United Arab Emirates (UAE) and Singapore, in partnership with Space Adventures. It is worth noting, however, that in the past a number of spaceports have been proposed only to falter for various reasons. A spaceport is a lot more than putting down a launch pad.

Full-Service Transportation
Overall, including the above list of US locations, there could be some three dozen operational spaceports spread out around the planet – most of them government owned and operated, but many of these newer projects are private initiatives. The promise of scheduled space liners blasting off with ticketed passengers is today just promise. In 2009 or 2010 this is scheduled to change, but even then we are talking about experimental flights where passengers have to sign all of their rights away to sue the Government and launch companies. In short "experimental passengers" blasting off from a spaceport must sign a statement that documents that they are aware that they may not come back.

"One might look at a spaceport as an innovative, new century version of what you remember airports first looked like," observed Patricia Grace Smith, formerly the head of the FAA's Office of Space Commercialization. "They will be a gathering place for people to learn and witness, for the first time, the capabilities and benefits of space."[3]

Smith pointed to New Mexico's spaceport intentions as "a full-service transportation entity for space. So you'll be able to go and take sub-orbital rides and experience zero-gravity, but also become educated and aware of all the various aspects of space."

Eric Anderson, president of Space Adventures, remarked that "... countries around the world are only just realizing the enormous commercial possibilities of space tourism." The market potential for sub-orbital space flights alone, Anderson has suggested, is estimated at $1-billion annually. If one truly plans and builds "space parks" along with the spaceports that number could mount into many billions.

Polarization of spaceport providers

Derek Webber, director of Spaceport Associates in Washington DC, has taken a hard look at spaceport types. He makes the case that it is probably not a workable plan to attempt to cover all markets with a single spaceport.

"There is emerging a polarization of spaceport providers," Webber observed. "Throughout the world, the already established government spaceports are likely to continue to provide expendable launch vehicle services to government, military and some commercial users. Meanwhile, new commercial spaceports are emerging that will focus primarily on space tourism–both sub-orbital and orbital–and will thereby support the development of the reusable launch vehicle mode of space flight."[4]

It seems unlikely, according to Webber, that a single, all-inclusive type of spaceport will emerge that is able to handle satisfactorily all the diverse kinds of spaceport business. This observation must be taken with at least a grain of salt in that Webber is in the business of promoting new spaceports.

In the spaceport business almost everyone involved has a separate view of how this will all work. Some believe that the government must operate launch safety range operations and be able to push a button to "terminate" a flight out of control. Others believe that private enterprise will take over everything and work much like a private airport. The Private Spaceflight Federation has no compunction in telling government officials that might slow down their flight to the skies to stay out of their way.

Fields of Dreams and the Fantasy Traffic Models

A more cautious and realistic perspective comes from Thomas Matula of the School of Business at the University of Houston – Victoria. Anybody engaged in the new spaceport boom should learn the lessons from the first one, he said. Speaking at the International Space Development Conference in May 2006, Matula explained that the first wave of spaceports occurred in 1989-1999. Those "fields of dreams", he said, were stirred up by such government projects as the Delta Clipper-Experimental (DC-X), the failed NASA / industry single-stage-to-orbit VentureStar program, and the privately-backed Kistler rocket.

Eventually, reality set in, Matula stated. Often spaceports focused on single firms, so when the firm failed so did the spaceport. Those backing spaceports didn't ask the "hard" questions, he said, like what is the real demand for launch services? Also, did the launch firms really have viable business models? And are the proposed launch vehicles technically feasible?

Matula said his counsel to spaceport proponents is that they must craft a realistic business model this time around. He suggested that most spaceports should at first serve as business incubators, not transportation facilities. The Mojave Air and Space Port that supports Burt Rutan's various enterprises, plus a growing number of other aerospace and space plane ventures, is a practical example of an "incubator" model. The problem is that this "model" has also proved to have its dangers. After the accident with the rocket test explosion that killed three people and injured three others, the US Government safety group known as OSHA, imposed a large fine for unsafe practices on Scaled Composites. In Mojave, the distinction between Scaled Composites, XCOR and the Mojave Spaceport is sometimes hard to distinguish.[5]

Thus Matula counsels most spaceport projects to start small and expand as needed, leveraging existing facilities before building new ones. He emphasized the need to ask hard questions about markets, revenues and viability of launch firms sooner rather than later. Of course not everyone is listening. There is a mega project in New Mexico, backed by Virgin Galactic, that is investing a quarter of a billion dollars in a state of the art facility. Clearly, Space Adventures also has the bit between its teeth and they and their partners believe that their entertainment-oriented business model can and will work.

Mojave Spaceport Makes the Headlines

Burt Rutan, the designer of the SpaceShipOne and Two – and other exotic aircraft – is the prime reason that Mojave Spaceport in California has gained a good deal of fame in the space commercialization world. This facility in California was the first inland commercial launch site to be licensed by the FAA. The license was awarded on June 17, 2004, allowing the airport – its full name is now the Mojave Air and Space Port – to support sub-orbital launches of reusable launch vehicles, that may include SpaceShipTwo and XCOR Lynx flights as well as other projects.

On the 100th anniversary of the Wright Brothers' first powered flight, December 17, 2003, Rutan's Scaled Composites, LLC, flew SpaceShipOne from Mojave Airport, breaking the speed of sound in the first manned supersonic flight by an aircraft developed privately by a small company. SpaceShipOne then flew from Mojave, past the boundary of space, fully loaded to meet the Ansari X Prize qualifications. On September 29 and again on October 4, 2004, Brian Binnie piloted SpaceShipOne to 112-kilometers (69-miles), to win the $10-million Ansari X Prize and at the same time smash the 107,960-meter (354,200-foot) altitude record set by the X-15 airplane in the 1960s. The landing was at the nearby Edwards Air Force base. At the time the sign that proclaimed "SpaceShipOne –NASA Zero" got a great deal of press coverage and discussion in Congress on Capitol Hill.

Rutan has predicted big things ahead for the Mojave Spaceport. As reported in the Mojave Desert News, Rutan said that "significant infrastructure" will be erected at Mojave to handle the space tourism business, including new space liner assembly facilities to be built within the next few months. He went on to say:

> "Oddly, spaceports are popping up but they have nothing to fly, with investments spurred by very little information. It's almost humorous to watch the worldwide battle of the spaceports … they're everywhere."[6]

The original Mojave Airport was established in 1935 and serves as a Civilian Flight Test Center, as the location of the National Test Pilot School, and as a base for modifying major military jets and civilian aircraft. One of the three runways is being extended as a critical element of the spaceport's expansion program aimed at the recovery of horizontal landing reusable launch vehicles (RLVs). The Mojave Airport is also home to several industrial operations, such as BAe Systems, Fiberset, Scaled Composites, AVTEL, XCOR Aerospace, Orbital Sciences Corporation, Inter Orbital Services (IOS), and General Electric. XCOR Aerospace has performed flight tests at this facility, including multiple tests with their EZ-Rocket which was successfully demonstrated in October 2005 at the Countdown to the X Prize Cup event in Las Cruces, New Mexico. In December 2005, the EZ-Rocket made a record-setting point-to-point flight, departing from the Mojave Airport, and gliding to a touchdown at a neighboring airport in California City.

New Mexico Chosen by Virgin Galactic

All the buzz and pizzaz about spaceports is not just smoke and mirrors for customer brochures. Virgin Galactic has now unveiled actual architectural plans for their Spaceport America in New Mexico. Clearly these designs suggest a sense of excitement and transports systems of the future."

A team of US and British architects and designers is developing 'Spaceport America' – a structure which they say symbolizes the world's first purpose-built commercial spaceport. The 100,000 square-foot (9,290 square-meter) facility is projected to cost about $31-million and will serve as the primary operating base for Sir Richard Branson's Virgin Galactic sub-orbital services, including the two White Knight Two carrier aircraft and five SpaceShipTwo spacecraft now under construction by Scaled Composites in Mojave, California. It will also provide the headquarters for the New Mexico Spaceport Authority.

This development follows on from the announcement made late in 2006 by New Mexico Governor, Bill Richardson, and Sir Richard Branson of Virgin Galactic to start operations from the South West Regional Spaceport in 2009/2010. The chosen location is some 45 miles north of Las Cruces and 30 miles east of Truth or Consequence – an area near Upham selected due to its low population density, uncongested airspace, and high elevation.

According to New Mexico planners, their eagerness to build the Southwest Regional Spaceport is driven by several money-making activities, such as:

- The emerging commercial space tourism sector, including operations of Virgin Galactic;
- NASA contracts for International Space Station commercial cargo and crew re-supply services;

Figure 5.3: The design for 'America's Spaceport' in New Mexico.
(Artist's Impression, Courtesy of Virgin Galactic)

- Proposed low-altitude racing competitions, such as those sponsored by the Rocket Racing League; and
- Evolving demand for low-cost human-rated reusable launch vehicles and rocket-powered racing aircraft.

According to Space.com, state officials have estimated that the economic impact on the region could be sizable. One probably wildly optimistic estimate has suggested that the Southwest Regional Spaceport could add to the region in excess of $750-million in total revenues and perhaps 3,500 new jobs by 2020. This "guess-timate" includes all commercial space transportation services, related manufacturing and services activities, plus tourism-related visitor spending.[7]

Another New Mexico client is the Connecticut-based UP Aerospace. According to the company's president, Jerry Larson, in an interview with Space.com: "Our site activation process is nearing completion at the temporary launch complex in New Mexico. Everything is progressing smoothly." UP Aerospace is planning an inaugural sub-orbital rocket blastoff from the New Mexico spaceport in the near future. "Our customers are performing final checkouts of their payloads and experiments prior to their integration with the vehicle," Larson said. "We're right on schedule for the multiple space launches that we have scheduled for this year."[8]

The Mid-Atlantic Regional Spaceport
The Mid-Atlantic Regional Spaceport (MARS) began with the creation of the Center for Commercial Space Infrastructure by the Old Dominion University of Virginia. This entity was established by the University in order to create a commercial space research and operations facilities within the state and it has been a prime mover in the development of a commercial launch infrastructure at Wallops Island, Virginia.

In 1995, the organization became the Virginia Commercial Space Flight Authority which has focused its efforts on promoting the growth of aerospace business in the region while also developing a commercial launch capability.

In July 2003 a bi-state agreement was reached between Virginia and Maryland to allow cooperation between the two states for future development and operations, and to promote the further development of the launch facility. Maryland now provides funding and the name was changed from the Virginia Space Flight Center to the Mid-Atlantic Regional Spaceport (MARS).

The FAA's launch site operator's license for MARS was renewed in December 2002 for five years and recently renewed again. The facility is designed to provide what its backers characterize as "one-stop shopping" for space-launch facilities and services for commercial, government, scientific, and academic users, although its targeted market is largely geared toward smaller payload missions.

In 1997, the original group signed an agreement with NASA to use the Wallops Island facilities in support of commercial launches. This 30-year agreement includes access to NASA's payload integration, launch operations, and monitoring facilities on a non-interference, cost reimbursement basis. There is currently a partnership agreement with DynSpace Corporation, a Computer Sciences Corporation company, of Reston, Virginia, to operate the spaceport. The State of Virginia, however, maintains ownership of the spaceport's assets.

Figure 5.4: Ready to launch at Wallops Island spaceport facility
(Photo source: MARS).

MARS has two launch pads at Wallops. The first, launch pad 0B, was designed as a "universal launch pad," capable of supporting a variety of small and medium expendable launch vehicles (ELV's). In March 2000, MARS acquired a second launch pad 0A originally built in 1994 for the Conestoga launch vehicle, which made one launch in October 1995 but failed to place its payload into orbit.

From its location on the Atlantic coast, MARS can accommodate a wide range of orbital inclinations and launch azimuths. Future plans include supporting reusable launch vehicles (RLVs). MARS also provides an extensive array of services including the provision of supplies and consumables to support launch operations, facility scheduling, maintenance, inspection to ensure timely and safe ground processing and launch operations, and coordination with NASA on behalf of its customers.

The successful launch of the four-stage Minotaur I rocket by MARS in December 2006 could be the beginning of a more intensive effort to encourage private commercial launch firms, such as SpaceX and OSC to look to the Virginia Eastern Shore's alternate spaceports. This could even begin to set the Wallops spaceport on a trajectory to send spacecraft to re-supply the International Space Station. With this aim in view, the spaceport began a $500,000, federally funded study to determine if it is a suitable location for orbital taxi missions to the space station sometime in the next decade. MARS is also ready to consider human sub-orbital flights too.

And on the West Coast

California Spaceport is co-located with the Vandenberg launch facility, operated by the US Air Force. This facility was the first commercial spaceport to be licensed by the FAA / AST, on September 19, 1996. In June 2001, the spaceport's license was renewed for another five years and a further renewal was granted as of 2006.

This facility offers commercial launch and payload processing services and is operated and managed by Spaceport Systems International (SSI), a limited partnership of ITT Federal Service Corporation. It is located on the central California coast and is able to leverage off the infrastructure developed to support the Vandenberg launch facility operated by the US Government. SSI signed a 25-year lease in 1995 for the 0.44 square kilometers (0.17 square miles) site.

The California Spaceport can support launches with azimuths ranging from 220 degrees to 165 degrees. Construction of the commercial facility began in 1995 and was completed in 1999. The design concept provides the launch pad with the flexibility to accommodate a variety of launch systems including low-polar-orbit inclinations. Although the facility is configured to support solid-propellant vehicles, plans are underway to equip the launch facility with the support systems required by liquid-fueled boosters.

Alaska's Kodiak Launch Complex

The Alaska state legislature passed legislation a quarter of a century ago in 1991 to create the Alaska Aerospace Development Corporation (AADC). It was structured as a public company to develop aerospace related economic, technical, and educational opportunities for the state of Alaska.

In 2000, the AADC completed the $40-million, two-year construction of the launch complex at Narrow Cape on Kodiak Island. It was the first licensed launch site not co-located with a federal facility and also the first new US launch site built since the 1960s. Owned by the state of Alaska and operated by the AADC, the Kodiak Launch Complex received initial funding from the USAF, US Army, NASA, the state of Alaska, and private firms. Today, it is self-sustaining through launch revenues and receives no state funding.

Kodiak is designed to serve several markets. These include military launches, government and commercial telecommunications satellites, remote sensing, and space science payloads weighing up to 1,000-kilograms (2,200-pounds). These payloads can be delivered into low earth orbits (LEOs), polar orbits, and highly elliptical orbits. The first launch was a sub-orbital vehicle, Atmospheric Interceptor Technology 1, built by Orbital Sciences Corporation for the USAF in November 1998. A second launch followed in September 1999. Further launches included a Quick Reaction Launch Vehicle, a joint NASA-Lockheed Martin Astronautics mission on an Athena 1 (the first orbital launch from Kodiak), in September 2001, and later, a Strategic Target System vehicle was launched.

In February, 2005, the Missile Defense Agency (MDA) launched the IFT-14 target missile, one of several rockets from Kodiak to test the US missile defense system. The AADC is also supporting development of ground station facilities near Fairbanks, Alaska, in cooperation with several commercial remote-sensing companies. The high-latitude location makes the Fairbanks site favorable for polar-orbiting satellites, which typically pass above Fairbanks several times daily

Spaceport Florida

In 1989 the State of Florida established the Spaceport Florida Authority and then renamed it in January 2002 as the Florida Space Authority (FSA). This entity is authorized by the State government to act just like an airport authority, to oversee the space launch industry and space-related economic, industrial, research and academic activities. The FSA occupies and operates space transportation-related facilities at the Cape Canaveral Air Force Station, owned by the US Air Force. FAA / AST first issued a license for spaceport operations on May 22, 1997, and renewed the license in 2002 and 2007 for additional five-year terms.

Under an arrangement between the federal government and FSA, excess facilities at Cape Canaveral have been licensed to the FSA for use by commercial launch service providers on a dual-use, non-interference basis. To date, FSA has invested over $500-million in new space industry infrastructure development in Florida. This includes a reusable launch vehicle (RLV) support complex (adjacent to the Shuttle Landing Facility at KSC), and a new space operations support complex. FSA is planning to develop an innovative, flexible and cost-friendly commercial spaceport to attract commercial launch companies to the state. This is also intended to accommodate the growing need for rapid response launch vehicles and the launching of smaller payloads for government, commercial and academic users.

A 2005 analysis by FSA included a market assessment of the number and types of launch vehicles that could possibly use such a facility, and concluded that a new commercial spaceport is feasible from both a market and technical standpoint. The conclusion reached by this study was that a Florida commercial spaceport would primarily benefit from the sub-orbital space tourism market – at least from an economic and usage level. It was estimated from this study that such a facility would generate increased economic activity, earnings, and jobs, and raise Florida's profile as a space state.

FSA is investigating the possibility of having specific Florida airports apply for an FAA Launch Operators license to support horizontally-launched spacecraft. They believe this will attract space tourism companies utilizing this technology, such as Richard Branson's Virgin Galactic. (In this particular case, however, Virgin Galactic has already signed a contract with the New Mexico Spaceport facility.) Several statewide airports have shown great interest in participating in space tourism but so far none has taken final action and sought FAA licensing.

Non-Federal US Spaceports

Several states in the US now plan to develop private spaceports that are intended to provide multiple launch and landing services. These proposed private spaceports have several common features. One unusual aspect of these projects is that rather than being coastal sites, whereby aborted missions can be easily accommodated, there are several inland facilities.

The Mojave spaceport was the only US FAA-licensed spaceport with an inland location rather than an ocean-based abort capability. However, some of these sites are generally designed to have sufficient space

Spaceport Name	Location	Owner & Operator	Launch Facilities	Status
Gulf Coast Regional Spaceport	Brazoria County, Texas (50 miles or 80 km south of Houston, Tx)	Gulf Coast Regional Spaceport Development Corporation	Launch Control Facility, Sub-Orbital Launch Platform	FAA Sub-orbital launch site licensing underway.
Oklahoma Spaceport	Washita County, Oklahoma (Burns Flat, Oklahoma – former Clinton-Sherman Air Force Base)	Oklahoma Space Industry Development Authority	13,500 ft runway, control tower, 50,000 sq. ft. manufacturing facility, maintenance & painting hangar,	OSIDA is conducting a safety and environmental impact study
South Texas Spaceport	Willacy County, Texas	To be determined	To be determined	This the final site to be selected by the Texas legislature. 3 sites are under study.
Spaceport Alabama	Baldwin County, Alabama (near to Mobile, Alabama but still it is an inland site.)	To be determined	To be determined	Phase 2 master site plan complete. Sub-orbital flight licensing process started with FAA.
Spaceport Washington	Grant County International Airport.	Port Authority of Moses Lake	Major runway and minor crosswinds runway. Emergency landing site of the Space Shuttle. 30,000 acre site.	Additional development is contingent on potential launch customers and financing. To be a vertical launch site
West Texas Spaceport	Geasewood, Pecos County, Texas	Pecos County, West Texas Spaceport Development Corporation	Control center, industrial strength concrete pad, 5 runways, extensive hangar space.	Seed funding from the State of Texas Plans for new runway, static engine test facility, & balloon hangar.
Wisconsin Spaceport	Sheboygan, Wisconsin	City of Sheboygan/ Rockets for Schools-Wisconsin Aeropsace Authority	Vertical pad for sub-orbital launches. Portable launch facilities and mission control.	Planning for additional facilities and devel-opment capabilities to support sub-orbital and orbital flights.
New Shepard Blue Origin Spaceport	Culbertson County near New Mexico and Texas Border	Completely private facility of the Blue Origin Corporation that supports the vertical takeoff & landing of the New Shepard vehicle	Launch pad, manufacturing and testing facilities. (Licensed by FAA for test launches)	This launch and test facility is now fully deployed in North Texas very near the New Mexico and Texas border. Test flight in Nov. 2006.

Figure 5.5: Status of Various US Non-Federal Commercial Space Ports[9]

and area size to respond to launch or landing mishaps and are often co-located with larger federal sites or former airport facilities. These projects, although private and commercially organized, are typically carried out in cooperation with state governments that are providing substantial incentives and in-kind resources to encourage spaceport development.

Assessment of US Spaceport Safety

The licensing of spaceport facilities, just as the flight worthiness of space planes and reusable launch vehicles, should remain on a case-by-case basis for some time to come. This is because the number of these facilities remains reasonably small and the nature of the launch operations at each site remains quite different. Certainly there are a number of key questions and issues that should be addressed in the case of each spaceport by FAA inspectors. The answers to these questions will then trigger further investigation and inspections. Different launch operations, different propulsion systems and different ground support requirements will clearly alter the nature and the stringency of the licensing process that follows. Key questions for spaceport licensing include:

a. **Abort and Mission Mishap Processes.** The licensing process must address the degree to which the launch site or spaceport has a large and securable perimeter area (and/or ocean frontage) that allows launch aborts or landing mishaps to be addressed with a minimum of safety risk to ground crew and surrounding residents and infrastructure. Land-based spaceports and ocean-adjacent sites give rise to different assessment issues. (Requirements as to who oversees and executes launch range safety and whether there are termination capabilities still need to be addressed.)

b. **Take-off and landing processes for space planes or re-usable launch vehicles.** Vertical take-off, horizontal take-off, jet carrier, jet-towed launcher systems and balloon-based launch systems all lead to different types of safety questions and regulatory concerns. Systems that involve ocean-based launches and balloon-based launches where the crew is remotely located (or largely remotely located) from the launch create a significantly different environment from one where the launch take-off is from a spaceport with a significant number of employees and ground crew. This is particularly the case where a vertical rocket launch and/or landing are involved.

c. **Escape capability.** One of the key issues is the degree to which there is a launch-to-land capability for crew and passengers to escape via a separate module or capsule. A key question is the landing provisions of the escape system and whether it can make a runway landing, a parachute or parafoil terrestrial landing or whether a splash down landing is required.

d. **Re-entry mode and the thermal protection system (TPS).** A critical part of the any space plane sub-orbital mission or de-orbiting from low earth orbit is adequate thermal protection and the nature of required aero-braking, parabolic arcs and pull out profiles that are required. In this respect, safety margins against engine malfunctions are particularly key and mission profiles should ensure that any accident would prevent a crash landing over urban or heavily populated areas and ability to land in glide mode.

e. **Basic Infrastructure.** Any fully licensed spaceport will need to be well equipped with testing facilities, fully backed up communications, power, security provision, and other critical infrastructure. Fortunately, since many spaceports are augmentations of modern and fully equipped airports most of this key infrastructure is already there. In today's uncertain world there is a need for security barriers, monitored access for all personnel and customers, parking, transportation infrastructure, etc.

f. **Testing, Assembly, Training and Simulation Facilities.** The spaceport should be well equipped with testing, assembly, flight training and simulation facilities. Alternatively access to these facilities should be provided at other convenient and accessible locations. In light of the accident at the Mojave testing facilities, clear safety standards need to be established and enforced for rocket and hazardous material qualification. Training facilities need to be overseen for adequacy and safety as well.

These are only some of the basic questions that FAA inspectors consider in licensing of private spaceports. They will also consider the degree to which the operator of the facility is fully financially viable, bonded and insured against accidental loss and liability claims by a responsible governmental entity or private equity company. In cases where issues of full viability or completeness of spaceport infrastructure remain, it might be advisable to grant only a provisional license, pending resolution of any remaining questions or pending successful performance. Fortunately, a great deal of experience has been gained with regard to the licensing of federal launch centers over the years and even commercial spaceports have, in some instances, been inspected and licensed for periods that now exceed ten years. The recent trend toward rapid expansion of commercial spaceports suggests that vigilant inspection and licensing procedures need to remain in place.

International Spaceport Projects

There have been a number of national launch sites owned and operated around the world for a number of years and commercial spaceport projects seem to have mushroomed in the past year. There is a clear appeal to many countries of having at least one spaceport in their country to create an image of modernity and being a part of a totally new, and "sexy" industry. The appeal is sufficiently great that some governments have even offered financial incentives to locate a spaceport within its borders, just as some city and state governments in the US have offered tax and other incentive.

Many new initiatives to create commercial spaceports around the world have been announced. These include projects in Canada, as described below. The US-based Space Adventures company is involved in spaceport developments in the United Arab Emirates and Singapore. And in Australia, the Woomera launch center is being modernized and upgraded to attract US operators such as the DTI Corporation to use this revitalized spaceport. (Once Rocketplane-Kistler was to have been the prime client for Woomera, but this does not currently seem likely.)

Clearly there are many more international spaceports to follow. The following sections report on these four international spaceport initiatives, which are indicative of the range of ventures that might be anticipated in coming years.

Canada's Spaceport Ranges

The Canadian Forces Meaford Range facility near Cape Rich, just off Georgian Bay and near Barrie, Canada, was selected by the Canadian Arrow / Planetspace operation as an initial test site and proving ground for its manned sub-orbital spacecraft. It was considered a prime spot for a private space launch facility because of its restricted airspace and waterways. The Cape Rich peninsula stretches 2.5 miles (4 kilometers) out into Georgian Bay and thus is almost equivalent to a barge launch. This Canadian military site is 70 square kilometers (44 square miles) in size. Initial test launch operations have been held at the end of a peninsula spit for safety purposes.

This site has hosted engine test firings and escape system shakedowns for the Canadian Arrow launch vehicle. The Canadian Meaford Range officials agreed to allow Canadian Arrow to use the base's facilities on a needs basis for these initial tests. The project received authorization from the Canadian transportation agency, Transport Canada, for test flights that go beyond the current engine test firings, but it now seems that actual test and operational flights will be relocated to Cape Breton in Nova Scotia.

The latest development in the planning for permanent facilities for the Canadian Arrow / Planetspace launch operations, however, involves a new location in Nova Scotia. Arrangements have been made for the spaceport to occupy 1 square kilometer (300 acres) of land on Cape Breton with the Nova Scotia government also providing financial and tax incentives. In addition to testing facilities, launch pads and manufacturing facilities there will also be a new state-of-the-art training facility for astronauts.[10]

The Singapore and United Arab Emirates Projects

The highly entrepreneurial Space Adventures Inc., which brokered the private flights to the International Space Station and offers a wide range of space tourism options, is planning to expand its activities greatly in the next few years. These plans include developing and operating both space planes and spaceports, and ultimately "franchising" the business around the world. As described in chapters 3 and 4, they are currently well along in developing two new space planes. The Explorer spacecraft is being developed in partnership with the Ansari family's Texas-based Prodea Corporation and the Russian Federated Space Agency.[11] The other was to have been the Xerus vehicle as developed by XCOR Aerospace, but in March 2008, XCOR indicated that its smaller Lynx spaceplane capable of lower altitude flight would come first. These spacecraft are expected to be commercially available in the not too distant future and Space Adventures intends to offer space rides at new spaceport locations, not only in the US but also in the United Arab Emirates and in Singapore.

It was in February 2006 that CEO Eric Anderson announced plans for these two new spaceports near major airports. "These sites in the United Arab Emirates and in Singapore are just the initial steps [for private space flight]," he said.[12]

According to Space Adventures' press releases, an agreement is in place to construct the first of these, at a cost of $265-million, at the Ras Al-Khaimah International Airport in the northernmost of the seven emirates that constitute the United Arab Emirates (UAE). Sheikh Saud Bin Saqr Al Qasimi, crown prince of Ras Al-Khaimah, said in statement in February 2006 that they also looked forward to expanding operations elsewhere. Space Adventures' CEO Anderson, in announcing the project, said Crown Prince Al Qasimi had been extremely supportive and had personally invested $30-million.

Space Adventures has also reported that it is working with a consortium of investors in Singapore to develop Spaceport Singapore, a facility that will offer not only sub-orbital space flights, but also astronaut training, parabolic flights to simulate weightlessness, and other high-altitude attractions. Current plans call for this $115-million complex to be built near Singapore's Changi International Airport.

Figure 5.6: Architectural Rendering of the Singapore Spaceport
(Courtesy of Space Adventures)

Lim Neo Chian, Chief of the Singapore Tourism Board, expressed strong support of the project when it was announced in 2006 by saying: "With the proposed Spaceport Singapore, we now stand at the threshold of an unprecedented opportunity to launch into space practically from our own backyard."

In its announcement of the UAE and Singapore spaceport projects, Space Adventures indicated that these two initiatives do not rule out an American spaceport in the future, although several unrelated projects are already well underway in the US The Zero G parabolic rides to achieve weightlessness originate in Las Vegas and Florida and these locations would seem to be prime candidates for US sub-orbital flights.

Woomera, Australia
The Woomera, Australia launch facility dates back to the 1960s. This was the site selected by the United Kingdom for its launch center when it was developing the Blue Streak rocket as a part of the European effort to develop an integrated multi-stage rocket for the European Launcher Development Organization (ELDO). This facility is being updated and modernized to support the operations of DTI and Rocketplane-Kistler (now in doubt) at a cost of some $80-million. The clear atmospheric conditions that exist in Woomera allow launches at virtually all times and the vast open areas that surround the launch site allow the recovering of staged rockets from the K-1 launch. (See earlier discussion of Australian launch development for further details.)

Two Possible UK Locations

In the U.K., two military air stations are being considered as locations for new facilities to service commercial space flights in the future. These are at the RAF (Royal Air Force) base at St. Mawgan, near Newquay in Cornwall and the RAF base at Lossiemouth, near the town of Elgin in Scotland. Lossiemouth has the advantage of not having controlled air space issues that could hamper St. Mawgan, where restrictions exist because of flight paths above Cornwall that serve aircraft traveling to and from London and Continental Europe across the Atlantic.

The Science and Technology Select Committee of the British Parliament warned the government in a 2007 report that it must provide the U.K. with a coherent space strategy if it is not to be left behind by other countries. The committee said the lack of government support for early stage technology development already places the U.K. at a disadvantage.

A spokesman for the committee said:

> "On space tourism, such as those ventures planned by Richard Branson's Virgin Galactic, the committee takes a different view from the government and is excited by the potential afforded by sub-orbital travel and the rise of space tourism industry. The MPs say they do not believe that it should be the responsibility of the government to fund this work but developments in this area should be encouraged through appropriate regulation."[13]

Virgin Galactic has made it clear that the U.K. should encourage private sector investment in space activities through commercial incentives, possibly through monetary and fiscal policy. The company has suggested that the government "invest in public-private-partnership type arrangements for basic infrastructures, such as space ports, that can be shared with the private sector for commercial activities." Will Whitehorn, president of Virgin Galactic, which is developing the facility in New Mexico, said there was potential for creating similar spaceports in the U.K. at Lossiemouth or St. Mawgan "if there was a suitable financial and regulatory climate."

There will likely be several other spaceport facilities in Europe to support the European-developed space planes. A site in Sweden for flights to see the "Northern Lights" has been mentioned in press releases and sites in other parts of Europe are in the pipeline to support the space planes now being developed in France and Germany.

Nationally Operated Governmental Launch Sites Around the World

To complete the global picture, there are a number of launch sites around the world developed and operated by governmental agencies. Some of these, like Woomera in Australia described above, may well play a role in the future of private space flight.

Many of these launch facilities are, in fact, operated by national defense-related agencies. Until the advent of space commercialization there were only about two dozen major space launch sites around the world and these are shown on the map below, with the site locations in the code key.

AUSTRALIA – WOOMERA

As described earlier, the Woomera site is being up-dated to operate as a commercial launch center. The Australian Space Council and the Australian Space Research Institute operate these launch facilities, which also support operations for the United Kingdom. Capabilities to support the Rocketplane-Kistler and DTI projects are to be provided under contract. A goal for Australia is to secure a share of Asian space business in launches, small satellites and space-based services. Woomera's advantages include polar orbit access, a sparsely inhabited downrange area, existing infrastructure and largely cloud-free weather.

BRAZIL – ALCANTARA

The CLA Alcantara Launch Center is located on the Atlantic coast outside Sao Luis del Campos where the Brazilian space agency INPE has its headquarters. This facility was expanded to handle the VLS orbital launcher; a formal opening was held in Feb.-1990. Pads are also provided for the Sonda 3/4 sounding rockets, meteorological rockets and other science vehicles. The position near the equator offers 25% greater advantage than Cape Canaveral from the Earth's rotation with regard to launches into GEO orbit.

CHINA – JIUQUAN SATELLITE LAUNCH CENTER

This facility was built in the early 1960s in the Gobi desert, 1,600 km west of Beijing, north of Jiuquan

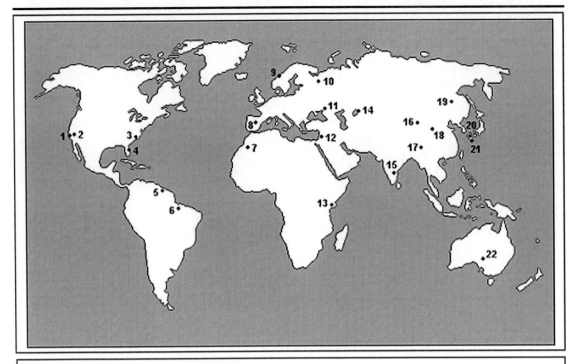

1 Vandenberg AFB, USA	13 San Marco Ocean Platform, Italy
2 Edwards AFB, USA	14 Baikonur / Tyuratam, CIS, Russia
3 Wallops Island, USA	15 Sriharikota (SHAR), India
4 Cape Canaveral / KSC, USA	16 Jiuquan, China
5 Kourou, Fr. Guiana	17 Xichang, China
6 Alcantara, Brazil	18 Taiyuan / Wuzjai, China
7 Hammaguir, Fr. Guiana	19 Svobodny, CIS, Russia
8 Torrejon AB, Spain	20 Kagoshima, Japan
9 Andoya, Norway	21 Tanegashima, Japan
10 Plesetsk, CIS / Russia	22 Woomera, Australia
11 Kapustin Yar, Turkey	
12 Palmachim / Yavne, Isreal	

city. Formerly known in the West as Shuang Cheng Tzu, this was China's first launch site but it is limited to SE launches into 57-70° orbits to avoid overflying Russia and Mongolia. Sounding rockets also utilize the site. Jiuquan continues to be used primarily for recoverable Earth observation / mg missions but because of its geographical constraints greater commercial activities are focused on the other two Chinese launch bases. A new launch pad to support a manned space project was built at Jiuquan for the Shengzhou program. In October 2003, China's first manned space flight was launched from Jiuquan, taking Yang Li Wei into orbit on board Shengzhu 5. This was followed two years later by the successful 5-day mission of Shengzhu 6, with a 2-man crew.

XICHANG SATELLITE LAUNCH CENTER
Xichang was selected from a short list of 16 sites from 81 surveyed for a more favorable GEO mission base than Jiuquan. Construction work 65 km north of Xichang city began in 1978 and resulted in the first launch in Jan. 1984. Xichang has two separate launch pads. The first can support up to five missions annually and the second complex is for vehicles with add-on boosters. The local population have traditionally been allowed to live close to the pads and when the maiden launch of the CZ-3B rocket crashed on 14 Feb. 1996 into a hillside 1.5 km from the complex, the death toll was six, plus 57 injured.

TAIYUAN SATELLITE LAUNCH CENTER
Taiyuan was inaugurated in Sept. 1988 for launches by CZ-4 into polar orbits for remote sensing, meteorological and reconnaissance missions. There is a single pad. It was initially operated for missile

testing as an extension of Jiuquan for larger vehicles, and launches began in 1996 carrying Iridium satellites. The site is designated Wuzhai by the US Space Command.

RUSSIA / COMMONWEALTH OF INDEPENDENT STATES (CIS) – BAIKONUR / TYURATAM

The Baikonur Cosmodrome offers nine launch complexes, with 15 pads, to accommodate space and missile activities. It is the only cosmodrome supporting Proton, Zenit, Energia and Tsyklon SL-11. Missile and rocket tests began in the area in the early 1950s, with construction of the pad from which both Sputnik 1 and Yuri Gagarin's first human space flight were launched starting in 1957. The site's location is actually nearest to Tyuratam in Kazakhstan, about 370 km SW of Baikonur. However, the Russians continue to give its latitude / longitude as that of the town of Baikonur. Kazakhstan finally named it after Tyuratam in 1992, but it is still widely known as Baikonur. Due-east launches (the most efficient) are not allowed from Baikonur because of the lower stages impacting in China.

KAPUSTIN YAR

Known as the Volgograd Station to its personnel, this was the first rocket development center, hosting its first launch in 1947. During its early years, Kapustin Yar tested captured V2 missiles and conducted sounding rocket experiments carrying dogs and other animals up to 500 km. Its first orbital launch was Cosmos 1 in 1962. By 1980 there had been 70 orbital launches, mostly small Cosmos science satellites, but as work was switched to Plesetsk the annual launch rate fell to an average of only one. Kapustin Yar in later years handled only occasional missions, possibly for radar calibration. There has been no orbital launch since 1987.

PLESETSK

Plesetsk was for a long time the world's busiest spaceport but has now been overtaken by Baikonur. Orbital launches have declined for a number of years, as programs are transferred to newer vehicles at Tyuratam. The base is located at 62.8°N / 40.1°E, which enables communications and spy satellites to be placed in polar and highly elliptical orbits. Construction of the first of several pads began in April, 1957, to support the nascent ICBM program. The pad and ICBM began active duty in January, 1960. There are nine operational pads: three for Cosmos, four for Soyuz / Molniya and two for Tsyklon.

SVOBODNY

Proposals for a new cosmodrome at the decommissioned missile site at Svobodny-18, 100 km from the Chinese border, surfaced during Russia's negotiations with Kazakhstan over Baikonur's future. They were widely perceived as a negotiating ploy, but President Yeltsin signed a decree on 1 March 1996 approving the cosmodrome's creation. Russia' successful Start booster was among the first users. However, during a visit to the Russian Far East at the end of September 2007, the first vice prime-minister, Sergei Ivanov, confirmed his government's decision to shut down launch operations in Svobodny, while reiterating intentions to find a new launch site in the region. Ivanov added that the new launch site in the Russian Far East would be available for manned launches among other missions. According to Ivanov, the government's strategic policy of bringing high-tech industries to the Russian Far East gave the region an advantage over other places in the selection process.[14]

In November 2007, the head of Roskosmos, Anatoly Perminov essentially ruled out Vanino from the list of potential locations, citing the area as earthquake-prone. He promised that the final selection would be made in 2008, adding that the Amur Region (Amurskaya Oblast) was now favored. According to Perminov, the new facility would only be necessary after 2020, upon conclusion of the Russian involvement in the International Space Station project, which is dependent on the launch site in Baikonur.[15]

FRANCE – CENTRE SPATIAL GUYANAIS (GUYANA)

Although owned by CNES, the Centre Spatial Guyanais (CSG) at Kourou is made available to ESA / Arianespace under a governmental agreement that guarantees access to the ESA-owned Ariane launch facilities. CSG is the most favorable major site for GEO launches, its near-equatorial position providing a 15% payload advantage over Cape Canaveral for eastward launches. Guiana's coastline unusually permits launches into both equatorial and polar / Sun-synchronous paths.

HAMMAGUIR (ALGERIA)

France built a launch facility in Algeria near the town of Mammaguir in the 1960s. There were, however, only four launches through 1985 and then the facility was closed. As the French Space Agency (CNES) moved its launch operations to Kourou these facilities were essentially abandoned as a base for French space launch services.

INDIA – SHAR CENTER
SHAR, encompassing the Sriharikota Island on the east coast of Andhra Pradesh, is ISRO's satellite launching base and additionally provides launch facilities for the full range of Rohini sounding rockets. The PSLV launch complex was commissioned in 1990. Modifications were subsequently made to allow the complex to handle the GSLV and other larger vehicles capable of placing satellites into GEO orbit.

ISRAEL – PALMACHIM
An area near or within Israel's Palmachim Air Force Base south of Tel Aviv, and near the town of Yavne, was inaugurated as an orbital site with the launch of the Ofeq 1 satellite on 19 Sept. 1988. Facilities are classified although they are visible from the coast road. Located at the eastern end of the Mediterranean, the site is restricted to retrograde launch operations for range safety considerations.

ITALY – SAN MARCO
Italy's offshore orbital launch platform is little used today. This facility is located in Formosa Bay, 4.8 km off the coast of Kenya but is owned and operated by the Italian government. It comprises two platforms. The San Marco platform provides the launch pad, while Santa Rita platform houses blockhouse facilities. The range is maintained in operational condition and the telemetry / tracking station still works under ESA contract. The range became operational in 1966. Eight satellites have been launched from this platform.

JAPAN – KAGOSHIMA SPACE CENTER
Japan's first six satellites were launched from these leveled hilltops facing the Pacific Ocean at Uchinoura on the southern tip of Kyushu Island. By the end of 1995, there had been 21 successful orbital launches, with 349 launches of all types. Construction of a sounding rocket site began in Feb. 1962, with extensions for science satellite launches by M rockets completed in 1966. The annual launch rate is typically 3-4 for the larger sounding rockets. Facilities have been upgraded to support the new M-5 orbital vehicle, first launched in Feb. 1997.

TANEGASHIMA SPACE CENTER
NASDA's Tanegashima Space Center (TNSC) orbital launch facility is located on the SE tip of Tanegashima Island, 1,000 km SW of Tokyo. The more northerly Osaki Launch Site includes the H2/J1 pads and liquid engine static test facilities, while the Takesaki Launch Site handles sounding rockets and provides facilities for H2 solid booster static firings and the H2 Range Control Center. A new pad was completed in 1999 for the upgraded H2 and further upgraded for the H2A launch vehicle.

UNITED STATES – CAPE CANAVERAL AIR STATION
Cape Canaveral Air Station (CCAS) is under the direction of the US Space Command 45th Space Wing. It encompasses active Titan, Atlas and Delta complexes in addition to providing support facilities for the military, NASA and commercial organizations. 455 space launches had been made by the end of 1995, including NASA's manned missions. However, the annual launch rate is now constrained to 25-30 because, despite upgrading, the Cape is based on 1950s concepts. Principal current launch activities are Titan operations at pads 40/41, Delta from Space Launch Complex 17A/B and Atlas Centaur from SLC 36A/B. Polar launches from Canaveral are not permitted, as this would involve overflight of populated land masses. CCAS operates the Eastern Range tracking network extending into the Indian Ocean, where it meets the Western Range system.

KENNEDY SPACE CENTER
Kennedy Space Center (KSC) is NASA's site for processing, launching and landing the Space Shuttle and its payloads, including Space Station components. Located on Merrit Island, adjacent to the USAF launch facilities of the Cape Canaveral Air Station, Kennedy was originally built to support the Apollo lunar landing program of the 1960s. After the last Apollo lunar launch in 1972, Launch Complex 39 supported Skylab 1973-74, Apollo-Soyuz in 1975 and Shuttle from the late 1970s.

POKER FLAT RESEARCH RANGE
Poker Flat Research Range (PFRR) is primarily a sounding rocket launch facility dedicated to auroral and middle to upper atmospheric research. Operated by the Geophysical Institute, it is the world's only university-owned launch range. It is also the only high latitude and auroral zone rocket launch facility on US soil.

VANDENBERG AIR FORCE BASE

Vandenberg Air Force Base (VAFB) is under the direction of the US Space Command 30th Space Wing. Vandenberg is responsible for missile and space launches on the west coast and operates the Western Range tracking network extending into the Indian Ocean, where it meets the Eastern Range system. Vandenberg provides the US with access into polar orbits using due south launches and was to have provided a base for Shuttle departures on high inclination missions. Vandenberg is also an important base for development and operational ICBM flight testing.

WALLOPS FLIGHT FACILITY

Now a part of Goddard Space Flight Center, Wallops became the third US orbital site in Feb. 1961 with the launch of the Explorer 9 balloon by the all-solid Scout. Nineteen vehicles achieved orbit from Wallops by end-1995 (plus three failures), the most recent in 1985. Although the facility is still available, Scout retired in 1994. Orbital activity resumed Oct.-1995, but the commercial Conestoga failed. Wallops' major activity now is as the base for NASA's sounding rocket program. Current vehicles include Super Arcas, Black Brant, Taurus-Tomahawk, Taurus-Orion and Terrier-Malemute. Some 30 launches are made annually.

WHITE SANDS MISSILE RANGE

White Sands is operated by the US Army and was the site of the first major US rocket firings after World War 2 before the move was made to Florida's larger range. It is also the site of the first atomic explosion. White Sands is still a major sounding rocket firing base and supports BMDO flight testing. White Sands Test Facility (WSTF) is operated by Johnson Space Center for Shuttle propulsion, power system and materials testing. WSTF is also responsible for White Sands Space Harbor (WSSH), which provides the third Shuttle landing site in the US after Florida and Edwards AFB.

References:
1. Space.com – http://www.space.com/businesstechnology/060517tech_spaceport.html
2. FAA/AST, "2006 Commercial Space Transportation Developments and Concepts: Vehicles, Technologies and Spaceports" FAA, Washington, D.C., January 2006
3. Leonard David, Senior Space Writer – space.com 17 May 2006
4. Spaceport Associates web site – www.spaceportassociates.com
5. Space.com – http://www.space.com/businesstechnology
6. Space.com – http://www.space.com/businesstechnology/060517tech_spaceport.html
7. Space.com – www.space.com/missionlaunches/050913_nm_spaceport.html; http://www.space.com/missionlaunches/050913_nm_spaceport.html
8. Space.com – www.space.com/news
9. FAA, 2007 US Commercial Space Transportation Developments and Concepts, Washington, D.C. 2007
10. Tariq Malik, "Private Space flight Group Selects Canadian Launch Site," Space.com, June 5, 2005 http://www.space.com/missionlaunches/050602_planetspace_launchsite.html
11. Alan Boyle, "New Group To Develop Passenger Spaceship" – MSNBC, February 17, 2006 http://www.msnbc.msn.com/id/11393569/
12. Space.com – Tariq Malik, Sub-orbital Fleet to Carry Tourist into Orbit in Style, Feb. 22, 2006 – http://www.space.com/businesstechnology/060222_techwed_spaceadventures.html
13. The Herald – http://www.theherald.co.uk/news/news/display.var.1549285.0.0.ph
14. AFP – http://www.space-travel.com/reports/Russia_To_Shut_Down_Svobodny_Space_Centre_999.html
15. AFP – http://www.space-travel.com/reports/Russia_To_Shut_Down_Svobodny_Space_Centre_999.html

— Chapter 6 —

"…I expect that over the next decade we'll see the price of seats drop from $200K to $100K to $50K, and perhaps as low as $25K per person. However, this all presumes we get up to a significant flight rate on the order of 5,000-10,000 sub-orbital passengers per year." – Peter Diamandis

BUT WILL IT WORK AS A BUSINESS?

The new promise of space travel becoming a private sector activity brings with it opportunities–and risks. Commercial space travel ventures will not be measured against the old metrics of national glory nor be sustained by government budgets and subsidies. Ultimately the normal business measures of profitability and return on capital investment will apply. One might argue that technological breakthroughs or advancement of society should be considered into the mix, but stockholders demand return above all – perhaps sometimes foolishly and shortsightedly so.

Furthermore, the commercial space travel market is a "thin" and high-end market. Unlike the dot.com on-line ventures of the 1990s where investors were willing to look at growth and wait years on profits, this is a much different kettle of fish – or at least serving a quite different group of potential buyers. The dot.com businesses were betting on a huge global market with billions of consumers ultimately buying "e-tail offerings." The number of potential commercial space travelers is hard to quantify at this time. You can bet your boots, however, that there are far fewer going to sign up for a space jaunt than those likely to buy a book or a CD from Amazon.com or BarnesandNoble.com. The big question of the day is whether this specialized market is perhaps a thousand times less or a million times less than the on-line consumer market. In this realm, like so many others, size really does matter.

The business entrepreneurs described in chapters 2 and 3 are well experienced in the harsh world of competition – and they have become rich enough to embark on yet another exciting new venture because they have been successful in the past. So far, none of them has made public their business plans and financial projections. But among all the speculation and guesswork, there are still a few clues about the realistic costs of the enterprise. Several organizations, including the space agencies, have also attempted to study and project the market of space tourism as a serious business. The bottom line is that no one will know how successful this new business will be until after the business has been launched in a serious way. Fortunately for the fledging industry, some very "deep-pocketed" billionaires are involved.

Most people, especially when the offering means putting your life on the line, will take a serious wait-and-see attitude about any such new business. Early adopters are fewer in number when the ante might be eternity.

The highly-respected American space-related consulting firm Futron, of Bethesda, Maryland, has undertaken a serious effort to study this new space tourism market and to identify the major critical success factors that they believe will apply. The nuts and bolts of their analysis we will delve into in just a while. Despite some careful methodology and their now two-stage assessment of this new market, we are still just guessing. Not until after the first spike of citizen astronauts has ridden into space, and we are able to see who else will be brave enough to form the more serious "second wave" of travelers, can the viability of the business be assessed.

There were surveys taken by "*Space Future*," as far back as the early 1990s. They found that as many as 70 percent of us want to go into space. Almost half of the respondents in one survey said they would pay three months' salary to do so. The problem with such surveys is that the answers were given on the basis of hypothetical "play money." Respondents were not paying out of a real pay check, nor had going into space become rather commonplace with several thousands of people actually having gone there and back.

More recent forecasts from Futron have predicted that the commercial space travel industry would start to operate in a serious way in 2009 and could generate revenues of more than $1-billion per year by 2021. This Futron study, together with other relevant inputs from experts in the field, thus suggest that a substantial demand is potentially there. In truth, the opportunity to fly into space is coming and soon. But how many are willing to shell out somewhere between $100,000 and $250,000 for a less than two-hour ride into the dark sky in order to experience some four-minutes or so of weightlessness? Soon it will be possible for anyone with the will and the wherewithal to venture into space, just don't forget your Dramamine and your wallet – a rather thick wallet.

Some Clues to Costs and Revenues

A great deal of publicity has surrounded the plans of Virgin Galactic, founded by Virgin Group chairman Sir Richard Branson and Burt Rutan, designer of the Ansari X Prize winner SpaceShipOne. Virgin Galactic aims to begin offering suborbital flights to the public beginning in late 2009 or early 2010. Carolyn Wincer, head of astronaut sales for Virgin Galactic, told TechNewsWorld that the company has already booked more than 200 tickets, priced at $200,000 each, for its roughly two-hour flights up to an altitude of about 110 kilometers (or 70 miles) and back down again. She added: "Sub-orbital trips are just the beginning. If we can prove that commercial sub-orbital travel can be profitable, what will happen is that private sector money will start to come in. When that happens, technology advances and prices come down."[1]

Stephen Attenborough, head of astronaut relations at Virgin Galactic, said that while the initial cost of $200,000 was necessary to recover the many millions invested in the venture, over time the aim was to reduce the costs to make space travel affordable for the masses.[2]

Figure 6.1: Virgin Galactic's SpaceShipTwo with wings feathered
(Courtesy of Virgin Galactic)

Will Whitehorn, president of Virgin Galactic, contends that the company's business plan is predicated on breaking even in 2012, and then making serious money after that. He still argues this is possible even though costs have ballooned by $50- million over their initial financial plan.[3] He also claims they have just under $40-million in flight deposits with Virgin Galactic based on a ticket price of $200,000 per person for a two-hour flight. The flight will include only four to five minutes of weightlessness as the ship approaches the limits of the Earth's atmosphere at a height of about 100 kilometers.[4]

Branson himself has said that he expects up to 3,000 passengers will travel with Virgin Galactic in the first five years, raising some $600-million in revenues. Eventually he hopes the cost of the flights will fall to around $20,000. No doubt all this is factored into his business plan, but further details are elusive.

Another of the billionaire space entrepreneurs, Robert Bigelow, who made his money from the Budget Suites hotel chain, has reportedly spent $95-million of his own cash on the development of his inflatable space hotels for tourists, as described in chapter 2. And he stands ready to spend as much again. Speaking at the Space Foundation's National Space Symposium in April 2007, Bigelow said he expects to offer four week stays in low Earth orbit for $11,950,000 in 2007 dollars, or $14,950,000 in 2012 dollars. This will cover pre-flight training and transportation there and back. He kindly offered to any would-be space travelers that they could now place a fully refundable deposit of 10% of the cost of the trip now.[5]

As well as private individuals or groups, Bigelow is also targeting governments and firms. By 2012, he says, he will be able to lease space on his orbiting stations to what he calls Sovereign Clients. These are governments and industries ranging from biotechnology to automotive. They could lease sections of the space station for confidential research work. Bigelow's leasing prices for the station module are quoted at $88-million to lease a module for a full year; or to lease half a module for a month the cost is $4.5-million.

Another example of a firm quoting prices is PlanetSpace, of Canada, which is accepting reservations for suborbital flights priced at $250,000 each, including two weeks of training. They expect to fly 2,000 new astronauts in the first five years and generate revenue of $200-million in the fifth year. The company plans to perform the first test flight of its Silver Dart orbital vehicle in early 2009, according to Geoff Sheerin, the company's president and CEO.

"If the airline business had been left to the government to develop, it would probably cost $10,000 a pound to fly on an airplane," Sheerin said. "The step we really needed to take, and now NASA has sanctioned it, is to have private sector suppliers of space travel."[6]

The pricing has certainly become more interesting with Jeff Greason of XCOR Aerospace of Mojave, Calif., claiming that his Lynx spaceplane will fly up to 37- miles high (or nearly 60-kilometers) for only $100,000, and starting at about the same time as Virgin Galactic. Greason certainly changed the competitive dynamics by announcing his rocket-powered vehicle that is substantially smaller, slower and less expensive to build than any of those proposed by rivals. Greason indicated that his spaceplane will cost less than $10-million to build (versus the estimated cost of $50- million for Burt Rutan's SpaceShipTwo). The Lynx vehicle is intended to carry only a pilot and a single passenger at twice the speed of sound to about 37 miles above the earth. The entire outing, which would begin and end at a conventional airport and include about two minutes of sub-orbital zero gravity, would take less than half an hour.[7]

Figure 6.2: An artist's impression of XCOR's Lynx vehicle
(Courtesy of XCOR)

Of course, this is a significantly shorter trip – and only half the ticket price – than that envisioned by Virgin Galactic, which is a more powerful six-passenger craft designed to travel at about four times the speed of sound and to fly over 60-miles above the earth. The key may be whether passengers might consider the slower, lower and cheaper flight to be safer as well. These and other companies entering the commercial space and space tourism sector are led by serious-minded businessmen who, without doubt, have their experienced number crunchers in the back office, as well as the engineers and marketeers to make it happen. But the announcements, speeches and demonstrations are long on optimism and enthusiasm – and short on details of financial plans, break-even dates and returns on capital investment. There are few precedents in private industry on a comparable scale, so what can be learned from past experience?

The Suppliers of Space Planes and Spaceport Services

The commercial space business involves more than travel services to the public. There are those that will build and service the space planes; and others that will provide the spaceports from which the citizen astronauts will travel. The leading developer of spaceplanes at this point is Burt Rutan and his SpaceShip Corporation, with Paul Allen and Richard Branson as partners. With sales of these craft to both Virgin Galactic and Jeff Bezos's Blue Origin corporation, Rutan's company is off to a good start. He has rather optimistically stated he hopes to sell at least forty of these vehicles. At $50-million a throw that would be $2-billion dollars in the bank.

Two observations, from a business perspective, seem obvious when it comes to the supply of space planes. One straightforward conclusion is that there are way too many potential suppliers at this incipient stage of the space tourism business. The forty something companies listed in Appendix A will continue to be winnowed down in number over time. There are already more than two dozen failed enterprises (as listed in Appendix B) and that number will only increase. Economies of scale in production will drive down the cost of future space planes and this is likely be a critical success factor for the production companies. Only a handful of suppliers seem likely to be able to sustain the production of these sophisticated and expensive vehicles.

The second observation is that the expansion of the market for space planes would be a tremendous boost for the potential suppliers. The expansion of the supply of space planes for space tourism flights to the supply of craft for executive jet travel thus becomes a key element of business planning. To date only a few companies have expressed serious intent to supply supersonic jets for executive and VIP travel; but the size of this potential market and the on-going need for executive travel, as opposed to the largely one-time space trips for citizen astronauts, is a major consideration that manufacturers will undoubtedly consider in their future business plans. These decisions are nevertheless far from simple. Lockheed Martin, for example, has estimated that it will cost $2.5-billion to achieve the transition from prototype to large scale production of the QSS supersonic executive transport.

The business elements of space port development are perhaps even more difficult to project. The economics of these facilities may actually turn on issues of national prestige and the desire to develop new jobs and industry as much as the desire to generate profits from space flights. The chapter on spaceports explained how many countries and state governments are offering tax relief and other incentives to attract a spaceport and the new high tech jobs associated with such operations. In some business plans, a spaceport is seen as some sort of space museum and some form of advanced technology amusement park. The emir of Ras Al-Khaimah has put up substantial capital toward converting the Ras Al-Khaimah International Airport to become a space port in partnership with Space Adventures. It is difficult to say how quickly the projected $265-million in investment could be recouped. The same could be said about the spaceport being developed in New Mexico to support Virgin Galactic operations at a projected cost of a quarter billion dollars.

The Lesson of Concorde

The Anglo-French Concorde was primarily an inter-governmental project which produced technological advances and huge national prestige, but never recovered its development costs of some several billion dollars – at a time when a billion dollars actually was a huge amount of money.

This aircraft, developed jointly by the British Aircraft Corporation and Aerospatiale of France, became famous as the only commercial, supersonic passenger jet and provided services for more than 28 years. Two fleets of Concordes were operated – one by British Airways and the other by Air France – between

January 1976 and October 2003. Their scheduled services linked London and Paris with New York and Washington DC in just over three hours flying time. At times, other routes were operated between Paris and Rio de Janeiro (with a refueling stop in Dakar, West Africa) as well as various routes in the Asia-Pacific region.

With fares in the $5,000-10,000 range for about 100 seats and luxury services, the aircraft were usually fully booked, but even so the full capital and operating costs could not be recouped. Although it was technologically brilliant, Concorde was a commercial failure compared with, for example, the Boeing 747 operating in the same long-range commercial market.

Figure 6.3: The Anglo-French Concorde
(Photo courtesy of British Airways)

The United States canceled its own supersonic (SST) program in 1971. Two designs had been submitted – the Lockheed L-2000, looking like a scaled-up Concorde, and the Boeing 2707, which was intended to be faster, carry 300 passengers, and featuring a swing-wing design. The Soviet Union also developed a competitor, the Tupolev TU-144, but the first production aircraft crashed at the Paris Air Show in 1973. The TU-144A (known as Konkordski) went on to make 102 scheduled flights between 1975 and 1978, when it was grounded after another crash.

Factors that limited the financial viability of both supersonic ventures included the limited range (for example they were unable to fly from Europe to the Far East, Australia or California without a refueling stop) and objections in various countries to overflying their territory because of the "sonic boom" and environmental pollution. It was even suggested in France and the United Kingdom that part of the American opposition to Concorde on grounds of noise was in fact encouraged by nationalistic spite at not being able to produce a viable competitor. However, other countries, such as Malaysia, also ruled out Concorde supersonic overflights due to noise issues.

The Concorde was the safest airliner in the world, measured by passenger deaths per distance traveled. But its reputation never fully recovered from the tragic crash of Air France Flight 4590 at Gonesse, near Paris, shortly after its take-off on 25 July, 2000.

There was another problem with the Concorde that was not as well publicized, but was quite serious. This was the damage done to the protective ozone layer at the top of the stratosphere. A thin layer of ozone serves as a protective shield from the deadly radiation that comes from the sun every day. On some days, when space weather is bad and there are solar irruptions, the radiation is particularly lethal. Without this protective layer there is a very serious danger of mutation to human genes and indeed of all fauna. In short, without the ozone protection the human species either dies or become a race of mutants. Corporate profits are good, survival of the species is, however, a tad more important. Money and wealth does not count for much if there are no humans around to take advantage of it. It is thus an inconvenient truth (someone has used that phrase before) that we need to look at the so-called "ozone issue" very seriously before large numbers of commercial space travelers take to the skies.

Anyway, there were mounting environmental and ecological pressures on the Concorde to end its flights when the SST was indeed grounded for other reasons. It is our view that space planes and the damage they might do the ozone layer is a significant public issue that needs to be addressed before the space tourism business takes off. To date the FAA has addressed public safety issues, but the concerns about space planes, the ozone layer and even the proliferation of carbon-based pollutants must be addressed in a serious way as commercial space enterprises move forward. This is particularly true if space planes move beyond space tourism to regularly scheduled flights for passengers from point A to point B and the volume of traffic surges upward.

The 'Vomit Comet'

Those looking for evidence of passenger numbers and public interest in space flight are encouraged by the success of Space Adventures. But perhaps even more significant are the Zero Gravity flights for passengers of more modest means who are paying far less bucks to experience weightlessness. It is not just high profile customers, such as the famed physicist Stephen Hawking who flew into weighlessness in early 2007. The ticket price for Zero G flights is just under $4,000, and more than 2,500 people have flown so far.

The Zero Gravity company, now part of Space Adventures, uses a modified 35-passenger Boeing 727-200, sometimes affectionately referred to as the "vomit comet." Flying from the Shuttle Landing Facility at the Kennedy Space Center in Florida, it provides 90-minute flights similar to those conducted by NASA to train its astronauts. The flights reach altitudes of between 24,000 and 32,000 feet, and passengers experience weightlessness for about 25 seconds on each parabola flown. Future flights from Las Vegas are also planned.

Figure 6.4: Space Adventures CEO Eric Anderson experiences Zero-G
(Photo courtesy of Space Adventures)

Space Adventures' Eric Anderson and ZeroG's Peter Diamandis have cleverly recognized that the dynamic range of the consumer market that exists between those who might fork out around $4,000 for a ride on the "vomit comet" and $20 to 35-million for a flight to the International Space Station is huge. The market size of those who might go into a casino and bet $5 is a lot different from those who might go into a casino and buy it. Thus the Space Adventures' web site, for instance, shows a lot of other options that cover the range from the casino bettor to the casino buyer. These options include a ride on the supersonic Russian Foxbat jet plane for around $19,000, or perhaps a week of astronaut training at the Russian Star City including a high "g" ride on the centrifuge to emulate a lift off to the Moon, and so on.

In addition to these "low end bets," of course, Space Adventures is still successfully marketing its flights on the Russian Soyuz spacecraft to the International Space Station (ISS). Six paying passengers have made the trip so far, at a cost of some $20-to $25-million each, and more are lined up to go in the next year or two. Recently Space Adventures announced that the basic price has risen to $35-million and a space walk would be another $15-million.

It is indeed a significant part of the business planning process for those seeking to enter the commercial space transport industry to see that a wide range of options should be offered to broaden market opportunities. The alternatives for market development range from some version of a space amusement park where people spend perhaps hundreds of dollars on up to a sustained stay in space for tens of millions of dollars.

Some Industry Economics

X-Prize founder, Peter Diamandis, gave his views on the economics of the industry in an interview with the Wall Street Journal. He pointed out that the cost of operating a mature transportation system (car, train, plane) is at present typically three times the cost of the fuel. So an airplane that burns about $5,000 of fuel per hour will cost an airline about $15,000/hour for all of its costs (i.e. leases, personnel, insurance, etc.). These economics, he said, hold true for systems that are reusable and operate at high flight frequencies – i.e. they spend more time flying than in the hangar.[8]

He continued: "The cost of sub-orbital flight seats today are projected to be $100,000-$200,000 per person. If you assume that the average craft will carry about six passengers then the revenue per vehicle flight will be about $1-million. The cost of the fuel on a sub-orbital system is probably on the order of $50,000 (max.), meaning the cost of operating the system can perhaps get down to $150,000 per flight and then drop lower."

Peter Diamandis went on to say: "For these reasons I expect that over the next decade we'll see the price of seats drop from $200K to $100K to $50K, and perhaps as low as $25K per person. However, this all presumes we get up to a significant flight rate on the order of 5,000-10,000 sub-orbital passengers per year."

Since that statement, the announcement of Greason's XCOR Lynx vehicle, as described earlier, appears likely to change the cost and price curve even more rapidly.

Another perspective comes from Paul Czysz, professor emeritus of aerospace engineering at St. Louis University. "The pioneers in North America understood the risks of their venture, but they believed they could control them," he told TechNewsWorld. The real question is whether the pioneers of the commercial space transporation industry can ultimately "control the risk" or not?[9]

Similarly, the potential of space travel rests not so much on developing brand new technologies as it does on understanding what's needed to make it feasible commercially. "Technology from 1942 still works," Czysz said. "It's understanding the skills required and the belief in your convictions that you can do it."

The Futron Predictions

The latest Futron study projects that by 2021 more than 15,000 passengers could be flying on sub-orbital trips each year, representing revenues in excess of $700- million. Futron works closely with the Satellite Industry Association and agencies of the US government and they have now carried out two detailed studies in this field – the first was in 2002 and then they updated their market projections in the second half of 2006.

Futron's assessment indicates that public space travel (or "space tourism") is beginning to evolve from being a fringe market, struggling to be taken seriously. If Futron's studies are right, commercial space can and will become a much more substantial enterprise. Their study suggests that within a decade or so this business enterprise will transition from a curiosity to a business – with revenues that in time could be meaningfully discussed as a true industry.[10]

Space tourism has maintained a steady level of activity, thanks to the occasional flights by the first "citizen astronauts" on the Soyuz missions to the International Space Station (ISS), brokered by Space Adventures at a cost in the range of $20-25-million per flight. As a result, sub-orbital space tourism has generated high public interest because it appears to represent the future. Private astronauts tend to be seen as a new type of "global celebrity" because not only is it high risk but it is also extremely expensive to undertake.

During this period, media interest was heightened by the $10-million Ansari X Prize, won in October 2004 by Scaled Composites' SpaceShipOne, and by the parallel activities of companies such as Space X, Orbital Sciences Corporation, Space Adventures, and Virgin Galactic. The fact that a growing number of commercial ventures are now selling tickets for commercial sub-orbital flights, scheduled to begin around 2010, again at a very high fare structure, continues to generate public and media interest – including the cover of Time Magazine and the Wall Street Journal plus features in newspapers around the world.

Although there has been extensive popular media discussion of consumer demand for commercial sub-orbital space flights and much speculation as to whether such ventures could possibly succeed, there is very little rigorous data on the extent of the market for such services. So far the market size is five people who, between them, have spent over $100-million. This is a difficult sample from which to project future sales.

The First Study

In 2002, Futron published its *Space Tourism Market Study* report that was perhaps the first structured attempt by a professional organization to provide an in-depth assessment of the potential market.[11] The report provided a forecast of the demand for orbital and sub-orbital space tourism activities – not just the number of passengers expected to fly, but also the number who would be financially qualified and would potentially consider booking such a service if it were available at different price levels. This initial marketing analysis covered the period 2002 through 2021.

The 2002 forecast was based on a comprehensive survey undertaken by Futron and included the input from the polling firm Zogby International of a sampling of individuals with the means to pay for such flights. This analysis differed from previous surveys that were generally speculative and simply polled the public at large using self-selecting non-scientific means.

Zogby International personnel conducted these in-depth surveys on behalf of Futron in 2002. They surveyed 450 wealthy people – defined as those with an annual income of at least $250,000 or a net worth of at least $1-million (in 2002 dollars). These surveys sought to determine their interest in space tourism and probed their willingness to take part in such flights at a number of price points. The survey pool was selected to best represent those people with the means to purchase such flights.

The survey instrument was also carefully crafted to provide a realistic depiction of space tourism, including potential risks and other downsides, and respondents were queried on various related factors, including their perception of the risk of space flight and other activities, amount of money and time spent on vacations, and fitness levels.

The survey results were then used to devise a quasi-model that might be applied to estimate what fraction of the worldwide population of people who could conceivably afford a space tourism flight, would, in fact, actually book such a flight once available. Included in this assessment was a "pioneering" factor, which eliminates people in the later years of the demand forecast, but whose interest in flying in space was primarily to be among the first to go; that is, once space tourism is not so novel, these people would no longer be interested, and were thus removed from the forecast. (By the end of the forecast period, flights to low earth orbit may also be available and this may or may not erode the sub-orbital flight market.)

The final step in the forecast process was to create what the Futron analysts used as their so-called "market diffusion model." Market experience has shown that the adoption of new technological services, such as commercial aviation, typically follows an established pattern, popularly known as an S-curve. This model shows the nature of customer acceptance and use of the service over time. This model thus assumes that there will a first be slow and growing acceptance of the service as the market becomes familiar with the product. This will be followed by a period of accelerated adoption as the market embraces the product and safety is established. At the end of the cycle there will be a deceleration as the market nears a saturation point. This model over time looks very much like an "S" Curve if it were stretched out over time.

To model this phenomenon in commercial space travel, Futron applied a Fisher-Pry curve to the total potential demand pool for sub-orbital service, with an estimated 40-year time frame from start to full market maturity.

(Author's Note: Other than using what might be considered too low of a "wealth standard" for the interviewees, the methodology seemed in many ways to be reasonable. Perhaps more significantly, there does not seem to be any allowance in the market projections for a major accident – especially one that resulted in the death of passengers – that could slow or even cripple this fledgling industry unless new and alternative technology that promised greater safety could be developed and quickly implemented. One can only look to the experience with the Shuttle and imagine what Shuttle operations might have looked like if it had been a commercial operation that experienced both the Challenger and Columbia disasters.)

Four Years On

Since the original forecast was completed and published in 2002, there have been many major developments in the space tourism marketplace. These are principally in technical aspects of sub-orbital space tourism and highlighted by such developments as the NASA decision to select two private aerospace teams (i.e. Space X and Orbital Sciences Corporation) to provide Commercial Orbital Transportation Service (COTS) to the International Space Station (ISS). There has also been significant progress on the regulatory front with the FAA and Congress developing specific guidance for physical and health standards for flight crew and passengers, rule-making for licensing experimental launches, etc.

In 2006, therefore, Futron decided to update their market analysis for Space Tourism and made the following adjustments:

a. **Changed the start date of the market forecast from 2006 to 2008:** When the original study was performed in 2002, Futron chose 2006 as the most plausible start date for passenger sub-orbital flights, based on the progress made by the companies competing for the X-Prize or who had otherwise announced plans to enter the market. An assumption in that analysis was that whoever won the X-Prize would then put the winning vehicle into commercial service shortly after capturing the prize. As it turned out, the prize-winning vehicle, SpaceShipOne, was retired by Scaled Composites immediately after winning the prize, and the company decided instead to develop a larger derivative, SpaceShipTwo, under a contract with Virgin Galactic. Other ventures are either taking similar approaches, or have found that it is taking longer than expected to develop their vehicles. As a result, the first passenger flights of commercial sub-orbital vehicles are now anticipated around late 2009 or early 2010. One can only applaud an approach that suggests that the best approach is a safe and cautious one in starting a space tourism business. In any event, Futron accordingly changed the start date of its new study.

b. **Increased initial ticket prices from $100,000 to $200,000:** At the time of the original forecast the "going rate" for a sub-orbital space flight was $100,000, based on sales made by the space tourism operator Space Adventures and pronouncements by other companies. However, Virgin Galactic, which could be one of the first entrants into the passenger market, is currently selling seats at approximately $200,000 per person for its initial flights, and other companies have raised their ticket prices accordingly – some being as high as $250,000. The initial forecast had set ticket prices at $100,000 for the first five years, then gradually declining to $50,000 by 2021. The new forecast sets the ticket price at $200,000 for the first three years, then gradually declining, again to $50,000, by 2021. This change affects both the overall potential revenue for the industry as well as demand, since some potential passengers will not be able to afford, or be willing to pay, the higher initial ticket prices.

c. **Updated population wealth statistics:** The population and growth rates worldwide of high net worth individuals has also been updated in this forecast, using resources such as the Merrill Lynch / Capgemini *World Wealth Report 2005*. The growth in this category has recovered in the years since publication of the original study, generating a small increase in the target population over the forecast period.[12]

The forecast of sub-orbital space tourism passenger demand from both the original study as well as the latest revision are shown in the following charts. Passenger demand is lower in the new forecast, with a projected demand of just over 13,000 passengers/year as of year end 2020 (See Figure 6.6) as compared to a total of over 15,000 passengers/year by year end 2020 in the original study (See Figure 6.5).

Figure 6.7 below provides alternative projections based on whether the industry matures within 35 years, 40 years or 45 years. This exercise suggests that if the industry "matures" and the price of flights accordingly drops more quickly, and safety standards and performance are perceived to have improved, there might be well on the order of 20,000 flights per year by end of 2020. But if the maturation rate for the industry is on the order of 45 years, then there would be a total of about 10,000 space tourist passengers per year by the end of 2020.

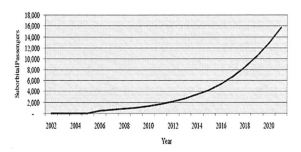

Figure 6.5: Original Estimate of Passenger Demand for Sub-orbital Space Tourism
(Courtesy of © Futron Corporation 2006)

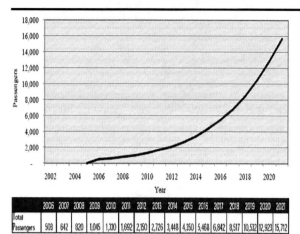

	2006	2007	2008	2009	2010	2011	2012	2013	2014	2015	2016	2017	2018	2019	2020	2021
Total Passengers	503	642	820	1,045	1,330	1,692	2,150	2,726	3,448	4,350	5,468	6,842	8,517	10,532	12,923	15,712

Figure 6.6: Updated Projected Demand:
Sub-Orbital Space Tourism
(Courtesy of © Futron Corporation 2006)

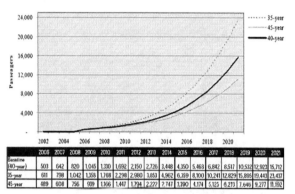

	2006	2007	2008	2009	2010	2011	2012	2013	2014	2015	2016	2017	2018	2019	2020	2021
Baseline (40-year)	503	642	820	1,045	1,330	1,692	2,150	2,726	3,448	4,350	5,468	6,842	8,517	10,532	12,923	15,712
35-year	611	798	1,042	1,358	1,768	2,298	2,980	3,853	4,962	6,359	8,100	10,241	12,829	15,895	19,443	23,437
45-year	489	608	756	939	1,166	1,447	1,794	2,227	2,747	3,390	4,174	5,125	6,273	7,646	9,277	11,192

Figure 6.7: Passenger Demand Forecast Using
Different Market Maturation Periods
(Courtesy of Futron Corporation 2006)

Of course there are many other factors that might significantly impact these forecasts. These sensitivity factors would most certainly include the number of companies that are offering space flights, the passenger capacity of the various vehicles, the degree to which this indeed becomes a truly competitive business, and the regulatory, insurance and "risk-management framework." The Futron study notes that one of the more significant factors may be how restrictive are the physical conditions set for passengers. There is the differentiation between passengers who remain strapped in for the entire flight versus those that might pay more for weightless activity. They note on the other hand that with the "strapped in for the duration of the flight" service, the health and physical condition standards for such flights might be reduced and that the potential market could accordingly increase.

The Disney Experience

Clearly, of all these factors, the "risk" or safety environment is likely be the most critical and any accident that led to the loss of life would have an immediate and major impact of the sub-orbital space tourism business, especially if there should more than one accident. Futron notes that in 2003, Walt Disney World's Epcot Center unveiled a new space-themed amusement park ride called "Mission: SPACE." The ride provides a simulation of space flight from launch to return, with a centrifuge providing the experience of twice the normal pull of gravity on the riders. Though this is lower in terms of G-forces than those experienced on other rides in the park, the centrifugal force is sustained for longer periods throughout the ride, as it would be on a space flight. This ride is clearly marked with signs indicating that any potential riders with blood pressure, heart, back or neck problems, motion sickness or other conditions that can be aggravated by this adventure should steer clear of the ride. Since the rides opening, there have been reports of health problems resulting from the ride, including two deaths within months of each other.

Since a sub-orbital flight may likely involve G-forces several times the force of gravity, health conditions may still be an important factor to consider and any thought of relaxing health and safety standards, at least until some years of experience has been gained, would be seem unwise. For this reason we believe that a more optimistic market projection made by Futron that the sub-orbital tourism market might actually rise to over $1-billion per year by 2021, if relaxed physical fitness standards were allowed, should be discounted at this time.

Finally there is the issue of how rapidly the sub-orbital space flight business might transition to actual orbital flights, space walks and access to commercially operated space habitats such as those which Bigelow Aerospace is developing. There are also other questions such as the amount of basic and applied research money that will be available to support the emergence of this new industry from governmental sources as well as the various high-flying entrepreneurs (in both senses of the words) who are supporting the development of this new industry. For example, the European Space Agency (ESA) has a program to provide finance and business advice to new companies, as described in chapter 4. The recent announcement by EADS at the Paris Air Show of its initiative to fly its space plane on commercial flights for space tourism has in part come from technical and institutional support provided by the ESA.

The Next Fifteen Years

After making all of the various adjustments to projected revenues, the initial total market forecast for the next fifteen years was revised downward by Futron from a high of $786-million/year to $676-million/year or a reduction of about 12 %. (The forecast is based on an initial price of $200,000 per person per flight, a 40 years time period for market maturation and the assumption that fairly high physical fitness standards are applied.) If this set of assumptions is taken as at least a reasonably carefully conceived forecast for this market, then the conclusion from this analysis is that this will not soon be a large global market in terms of revenues. This is particularly true if we compare these projections to the global aerospace manufacturing and global aviation services industries that represent over a trillion dollars per year in sales.

What is of particular note is that this is an industry that needs a tremendous amount of basic and applied research for it to succeed. These R&D costs are not only for the development of launch systems and cost-effective operations but also for safety-related research, space systems safety design and life-support suits. This industry, for some considerable time, will need to be heavily subsidized by governmental space agencies, such as NASA's $500-million commitment to support COTS commercial launch service to the ISS, as well as the entrepreneurial support from commercial leaders such as Sir Richard Branson, Paul Allen, Jeff Bezos, Robert Bigelow, John Carmack, Elon Musk and others.

The real question thus tends to become whether there is simply a large enough market for the sub-orbital tourist business to sustain the needed investment? It may be that the sub-orbital space tourism market is only a part – perhaps a small part – of a future space commercialisation market that can be sized in the billions of dollars a year that will be needed to make the necessary transport and safety research investments viable. Key among the various questions, therefore, is whether the commercial space market will at some point include sub-orbital transportation systems? In short, will new space plane systems also be used to accommodate business executive travel using sub-orbital trajectories?

References:
1. Katherine Noyes – *Techworld News* – 5/17/07
2. Travel & Tourism News – http://www.ttnworldwide.com
3. London *Daily Telegraph*, August 11 2008.
4. *SpaceNews.com*, April 1, 2008
5. Flight – http://www.flightglobal.com/articles/2007/04/10/213196/bigelow-aerospace
6. Katherine Noyes – *TechWorld News* – 5/17/07
7. *Economy Fare ($100,000) Lifts Space-Tourism Race,* New York Times -March 26, 2008
8. Wall Street Journal – http://forums.wsj.com
9. Katherine Noyes – *TechWorld News* – 5/17/07
10. Futron Corporation, *"Suborbital Space Tourism Demand Revisited"* – August 24, 2006 www.futron.com
11. Futron Corporation- *"Space Tourism Market Study, 2002,"* http://www.futron.com
12 . Futron Corporation – *"Sub-orbital Space Tourism Demand Revisited"* – August 24, 2006 www.futron.com

— Chapter 7 —

"Congress said in the Commercial Space Launch Amendments Act, our authorizing legislation, that "the future of the commercial human space flight industry will depend on its ability to continually improve its safety performance." There is no one that I know that has to be convinced of this" – Patricia Grace Smith (former US Federal Aviation Authority)

How Safe is Private Space Travel?

To encourage Citizen Astronauts to venture into space in significant numbers, it will be essential to create a sense of safety. Over time, this sense of reliability must increase to the level experienced in the commercial aviation industry. This safety factor is especially key if space tourism is to branch off into an entirely new commercial venture, namely hypersonic aviation transportation. This might be for commercial flights, such as those associated with the ESA-backed A-2 spaceplane that could fly 300 "space passengers" across the globe in 5 hours, or the Quiet Super Sonic business jets for CEOs and high flying billionaires – with less than a dozen high flyers aboard.

To achieve this new attitude towards the safety of space flight, the history of NASA's two tragic Shuttle disasters, will need to be consigned to history – in the same way as major airline crashes that have occurred around the world have eventually faded away. Today, anyone who looks seriously at the history of human space flight and related vital statistics would note that 1% of all such launches have ended with fatalities and even more disturbing 4% of all astronauts and cosmonauts (recognizing that most went on multiple missions) ended up rather terminally dead. Clearly we have a long way to go. One can only hope that commercial projects with corporate survival in mind will be able to improve reliability and success by many orders of magnitude.

The pioneering Citizen Astronauts that sign up to travel on space planes in the new, shining world of space tourism and, in time, hypersonic travel must be reassured that both the new space tourism industry and government regulators are focused on safety issues.

In the US, the Federal Aviation Authority's rules for commercial space flight are generally considered to be relatively light as far as regulations go. Applicants for experimental permits will need to provide a description of their program, a flight test plan, documentation showing the operational safety of the spacecraft and a plan for response in case of a mishap. Space vehicles are also designed to be able to handle an unlimited number of launches. Despite these regulations, the US Government has made it clear that they still consider these flights experimental and are not planning to "certify" spacecraft for some time to come.

The newly adopted FAA rules call for launch vehicle operators to provide certain safety-related information and identify what an operator must do to conduct a licensed launch with a human on board. The operators must also inform passengers of the risks of space travel in some detail. They also ask all passengers to sign a waiver to hold the US Government harmless if there should be an accident.

According to former FAA Administrator Marion Blakey:

> "...from the government's perspective, our official policy is this ... to embrace the private sector's daring spirit and clever ingenuity. And yes, you better believe that includes space tourism. We are in the business of encouraging and enabling the private sector. We develop regulations to make this high-risk business as safe as possible ... And we make sure potential passengers are properly informed and are willing to accept the risks that remain. And then? Well, then we'll step aside ... get out of your way ... and let you do what you do best: innovate."[1]

The US Government's position was further underlined in a Wall Street Journal interview in April 2007 with the headline: "How Safe Is the Race To Send Tourists into Space?" It was in the course of this interview that Patricia Grace Smith (then head of the FAA Office of Space Commercialisation and now on the Board of SpaceDev), said: "I believe that to work with industry to ensure safety is an appropriate governmental role. We are happy to play it given the fact that our industry is clearly on the same page. In giving us responsibility for regulating and promoting private human space flight, Congress directed us to protect the uninvolved public. The space flight participants, or passengers, travel at their own risk, once they have been provided the safety record and data related to the vehicle they are flying on. The space flight companies that we are working with are clearly focused on safety … and believe strongly that safety leads to reliability leads to more business and sustainability. So do we."[2]

Figure 7.1: former FAA Associate Administrator Patricia Grace Smith

In the Wall Street Journal interview, Ms. Smith went on to recall an FAA conference in February 2007. A panel was entitled, "When is a launch vehicle ready to carry passengers?" and industry leaders Alex Tai (Virgin Galactic), Jeff Greason (XCOR), George Whitesides (National Space Society) and John Herrington (Rocketplane-Kistler) all resoundingly stated that the "vehicle will fly when its safe to fly."

In the same Wall Street Journal interview, the space industry pioneer, Peter Diamandis, said: "I would actually love to see some of these companies strive for larger breakthroughs, but they will probably not do this in the early days because they need to strive for safety and a reliable revenue stream."

In another interview, Alex Tai, Virgin Galactic chief operating officer, said he believed the space tourism industry could bounce back from a disaster. He explained that the industry would need to act carefully and responsibly with passengers fully warned of risks before take off. If properly briefed on the potential risks and dangers of space flight, customers would likely be unable to file a lawsuit and win in case of an accident. "God forbid it should happen on the first flight. Hopefully it's many, many years out," he added.[3] Others from the legal profession who have taken on liability suits in the past are not so sure. Right now the one specific thing that is happening is that the states that are hosting spaceports are revising their state laws to line up with federal regulations in terms of liability provisions.

The Case for Health Testing

In the US, a leading medical establishment has suggested screening pilots and passengers for heart disease, balance disorders and other medical challenges posed by out-of-this-world travel. The Mayo Clinic in Arizona has teamed with a Southern California training lab and the University of Texas to provide such comprehensive testing and training for passengers, pilots and other commercial space travelers.[4]

"Many people have the perception that this (commercial space travel) is something that is completely innocuous," said Dr. Jan Stepanek, director of the Mayo Clinic Aerospace Medicine Program in Arizona. "This is not the same situation such as the NASA astronauts where you can select as many as you want and you end up with the fittest of the fit and the healthiest of the healthy. Many of these people (commercial space flight customers) may be in their 50s and they may have some health issues, cardiac issues."

Doctors at Mayo's Scottsdale campus already have performed medical checkups on NASA astronauts. Stepanek and another Mayo Arizona physician, Robert Orford, are board certified in Aerospace Medicine and Internal Medicine. "The most important thing is to maintain safety and provide patients with correct information," Stepanek said. "That would include what would be safe, unsafe and potential for injury."

Figure 7.2: The Mayo Clinic, Scottsville, Arizona
(Courtesy of Mayo Clinic)

Health Advice for Virgin Galactic Passengers

Sir Richard Branson's Virgin Galactic group has turned to a US company for specialist health and safety advice. Some 250 people, including 35 Britons, have so far paid a deposit to fly into space with Virgin Galactic and the company has given Wyle Life Sciences Group a contract to provide their passengers with advice, a chief medical officer, management services and analysis of medical data. As part of their preparations, many of the first one hundred of SpaceShip Two's passengers (including Branson himself) have already been through medical assessment and centrifuge training at the he National AeroSpace Training and Research Center (NASTAR) facility in Philadelphia.

Wyle is no newcomer to the world of space travel. For forty years the company has been working with NASA by helping with analysis, medical tests, training and other support services needed to prepare astronauts for space flights.

The position of chief medical officer falls to Wyle's Dr. James Vanderploeg and his responsibilities will include setting up requirements for training, medical checks and putting in place medical protocols. Speaking about his appointment, Dr. Vanderploeg said, "Supporting one of the leading proponents of private human space travel is both exciting and professionally rewarding. Wyle has the right blend of experience and knowledge to successfully support this rapidly maturing industry."

The chief operations officer of Virgin Galactic, Alex Tai, said of Wyle's support, "We are delighted to bring Wyle on board the Virgin Galactic team. It became obvious that they bring invaluable experience and resources into this important element of the program."[5]

The cost to be a passenger on board the first Virgin Galactic commercial space flight does not come cheap at $200,000 a ticket. This fare has already been fully paid by the 100 members of the 'Virgin Galactic Founders', who will board SpaceShip Two, hopefully, in 2009 or 2010.

Areas of Concern

The FAA regulations went through a lot of public scrutiny as the rule making process took place over a protracted discovery and hearing process in 2006 and 2007. But until there is actual experience with commercial passengers riding to the edge of space no one is sure whether the safety controls for the emerging new space tourism business are indeed adequate. Certainly areas of concern remain. The procedures for the notification of dangers that operators must follow with regard to potential "space tourists" seem to be reasonably stringent. But space plane operators do not have to report on the performance of "prototype and experimental" versions of the craft they operate. The process by which operators can shift a craft from experimental to operational status by renaming it or perhaps making only superficial changes in the design is an area of some concern.

Even more significant is a review of the formal rule making process that the FAA followed in creating the regulations now in place. On one hand it appears very clear from the record that the FAA considered very carefully the various interventions made by participants. In some cases, the FAA acknowledged the validity of industry interventions and made changes; and in other cases they rejected the arguments and suggested alternative language. Those who provided comments in the rule-making process were almost all from those seeking to operate space tourist businesses and develop space plane technology. There was an absence of individuals and organizations with the knowledge of this new technology to address proposed safety regulations from the perspective of the customer or the safety of the general public.

The FAA has been placed by the US Congress in a conflicted situation with its regulation of commercial space flights. On one hand the official Office of Space Transportation Report – "*2007 US Commercial Space Transportation Developments and Concepts: Vehicles, Technologies and Spaceports*" – indicates that its mission is to: "license and regulate" US commercial space activities and to "ensure public health and safety and the safety of property while protecting the national security and foreign policy interests of the United States during commercial launch and re-entry operation."[6] However, the FAA is also "directed to encourage, facilitate and promote commercial space launches and re-entries."

Thus the FAA has been directed by current US law to play the role of referee and umpire as well as cheerleader for this emerging new market. In the case of aviation it regulates and licenses, but the FAA is no longer charged with promotion of air transportation.

In short the FAA Office of Commercial Space Transportation, unlike the rest of the FAA, has been directed by legislation and Executive Order to promote the development of this new industry while also developing and enforcing safety regulations. Further it has been directed to await experience achieved through 2012 with regard to safety issues before making modifications in its safety regulations. It has, however, also been directed by Congress to hire outside contractors to study a number of issues, including the reasonableness of retaining both the licensing and regulation role on one hand and the promotion role on the other.

Experience with regard to NASA's role as the builder, operator and risk controller of the Space Shuttle, through its Office of Safety and Mission Assurance (OSMA) and Chief Engineer, suggests that conflicting priorities can be both real and deadly. In the case of NASA, however, this was not the fault of the OSMA

On more than one occasion, OSMA was overridden on critical safety issues during a full agency Shuttle readiness review. In many cases, OSMA opted to sign a statement of safety reservations; they were then overruled at a higher level of NASA management. Nevertheless OSMA was involved in reviewing and approving an excessive number of waivers – in some cases these waivers led directly back to factors that, in turn, led to deadly accidents.

Comparing NASA to the FAA

The roles and responsibilities of NASA and the FAA are in many ways quite different. The current duties of the FAA are largely of a regulatory nature. Its primary mission is to oversee the maintenance and improvement of safety in aviation. In the future it will also be oriented to oversee and improve safety in space tourism and other commercial space ventures. As such, it might be appropriate to rename the FAA, the Federal Aerospace Administration.

The FAA now has the new responsibility for overseeing the safety and reliability of high altitude platform systems and private space transportation systems and it appears possible that its domain will extend in time to low earth orbit and eventually all the way to geosynchronous orbit. From the regulatory perspective, the FAA has the ability to monitor key statistical elements of safety performance and risk mitigation. While the FAA operates tracking and traffic control networks and researches new and improved systems, it does not operate airlines nor develop new airplanes or space systems. These tasks are left to the operators of airlines or the manufacturers of aircraft and various types of airships.

NASA, on the other hand, has a wide range of duties and responsibilities. This includes R&D for aircraft, launch vehicles, satellites and new space exploration systems, operation of various types of these systems as well as operating networks to communicate, command and control these networks. It has a special challenge of not only developing and operating the most experimental and state-of-the-art systems but also maintaining the safety of their operation – as an operator as opposed to as a regulator.

The FAA has one great advantage when seeking to improve the safety of aircraft. When first licensed to operate, the aircraft has extensive flight experience and, over time, has accumulated tens of millions of hours of operation. NASA, on the other hand, is responsible for developing systems that have only minimal levels of testing and operational performance in comparison. This is simply because rocket launches are infrequent and last only a few minutes in duration in comparison to airline flights that are frequent and often last for many hours. The testing and certification of aircraft produced in volume to the same specifications is thus much easier to achieve than testing and certifying space vehicles.

NASA's attempts to improve astronaut safety are very heavily focused on developing new technology and new launch systems that can perform more reliably. The FAA safety-related processes are much more focused on management, performance goals, and operating processes, such as setting safety measures and objectives for aircraft, airports, weather information networks and operational personnel as well as upgrades to existing aircraft infrastructure. NASA is perhaps 80% focused on new technology and 20% on operating processes and performance goals for equipment and personnel, whereas the FAA might be said to be 20% focused on new technology and research and 80% on operating processes and performance goals. There may be logic in both agencies shifting toward a more balanced effort as between R&D / new technology on one hand and improved operating procedures and safety culture issues on the other.[7]

The New US Legislation on Private Space Ventures.

In December 2004, the US Congress passed a new law that gave the FAA additional authority over private space initiatives and this was signed into law in early 2005. Additional Congressional hearings with the FAA took place in February 2005.

Then in December, 2005 the FAA sent to Congress an extensive report outlining in great detail the rules and regulations that it intended to put into effect with regard to space tourism and other private space ventures involving the launch of "private vehicles" with people on board.

Despite the length of the proposed rules, the requirements are not particularly onerous. The operator of a space tourism business is only required to: (i) have a pilot with an FAA pilot certificate (Class two only) and each crew member would need a medical certificate issued within a year of the flight; (ii) there must be training of the crew and pilot to ensure that the vehicle would not do harm to the public even if abandoned; and (iii) operators of such systems must inform all participants of the associated risks and have them sign a consent form. Physical examinations for passengers would be recommended but not required unless a "clear public safety need is identified."

In terms of the licensing of "manned launch systems" intending to provide space tourists with access to space as well as the licensing of spaceports, the FAA has essentially indicated that each review process would be on a case-by-case basis. This is because each program and spaceport facility tends to be highly individualized to the extent that comprehensive guidelines are not possible.

The International Perspective

By the time that regular space tourist flights actually begin within the next one or two years, it seems essential to attempt to clarify international law and set minimal standards and registration procedures. There is no need for these to be overly restrictive and national governments (i.e. in the country where the commercial entity is registered) could and perhaps should continue to exercise most of the regulatory oversight.

As noted earlier and detailed in Appendix A, there are well over 40 entities worldwide engaged in trying to develop private commercial access to space at this time. Regulatory oversight would be largely to support this new and emerging space tourism market in terms of providing certification of minimal safety standards and to insure that nationally regulated programs do not interfere with or endanger one another at the global level.

In addition to the various space plane and "access to orbit" projects there are an even larger number of sites where spaceports are under development, with some already in operation. This number is continuing to grow and some form of structured control and regulation appears fully justified and necessary. However, no public international organization has so far been chartered with the responsibility to regulate the safe operation of launch vehicles, international space stations, or space exploratory missions.

One significant initiative was the formation in 2005 of the International Association for the Advancement of Space Safety (IAASS). This organization is sponsored by the European Space Agency (ESA), and the national space agencies of the USA. (NASA), Japan (JAXA), Russia (Roscom), Canada (CSA), France (CNES), Germany (DLR), China (CNSA) and Italy (ISA), among others. This entity is dedicated to the idea of creating international standards for space safety and is currently embarked on the development of new texts that record in a single place a great deal of stored wisdom and scientific data concerning space safety.

The IAASS has thus sought to strengthen professional training and academic study of space safety. It is also developing more effective and unified ways to manage safety on board the International Space Station. The IAASS-sponsored handbook on "Safety Design for Space Systems" is to be published in 2009. This will be the first attempt at creating a totally comprehensive text on space systems safety design as compiled by safety engineers and space systems practitioners from around the world. A follow-on IAASS text is also planned to look at innovations in the space safety field that will come with new technology, operational techniques and improved safety practices in future years and from the advent of private commercial space travel and new space plane systems.

IAASS has also undertaken a study of how commercial or private space initiatives and private space stations might be regulated and licensed at the international level. This study is being carried out in cooperation with the International Civil Aviation Organization (ICAO), the specialized UN Agency, headquartered in Montreal,

Canada, that is charged with the development of international standards for safety in aviation. The reasoning is that many of the legal and regulatory practices and approaches to safety that have been developed for global aviation might be productively applied to private space initiatives as well.

One suggestion that has been put forth is the creation an Independent Space Safety Board that would provide international safety certification services to the space-tourism industry. The advantage of this approach is that it would operate on a commercial (but non-profit) and presumably streamlined basis. Thus it would not involve the elaborate and often costly operations of a UN specialized agency. Further, this approach would presumably leave much of the regulatory responsibility with existing national regulatory agencies. Others have proposed that these responsibilities simply be formally assigned to ICAO.[8]

The Role of the International Telecommunication Union

At present, the United Nations has no specialized agency directly charged with regulating or setting international safety standards for space activities. Nevertheless there is some degree of international regulation and standardization already in place and there are several "space treaties" that have been signed by many space-faring nations that are currently in effect.

The International Telecommunication Union (ITU), headquartered in Geneva, Switzerland, in particular, has developed recommended standards for the launch, operation, and registration of applications satellites, space stations and other types of space systems in terms of registration, allocation and use of radio frequencies. These ITU regulatory processes also include "due diligence" activities carried out by national governments with regard to the safe launch of satellites. These procedures address, for instance, how satellites, space stations and space vehicles are to be de-orbited from low, medium or geosynchronous orbit.

These procedures also cover the review and inspection processes that governments are requested to carry out to ensure that the separation of various stages and components in a rocket launch are conducted so as to avoid the accumulation of space debris in orbit. The ITU, however, has no directly designated role in the regulation or control of space safety, except so far as it involves radio transmissions, although its recommended "due diligence" review and inspection procedures are generally observed around the world by current space-faring nations.[9]

Another UN-mandated agency, The Committee on the Peaceful Uses of Outer Space (COPUOS), headquartered in Vienna, Austria, considers the broader legal issues involving exploration and use of outer space. Their deliberations have helped to develop treaties and international conventions related to the exploration of outer space, including the international status of the Moon and other celestial bodies. However, it plays no direct role related to space safety and the reduction of risk related to space launches and space operations, except through its creation of a group to develop improved procedures to reduce space debris. The new procedures, that required unanimous consent took almost two decades to develop and many feel these new regulations add up to "too little and too late."

COPUOS is thus, in truth, simply a coordinating committee of the United Nations organization. Its current membership includes all space-faring nations as well as other nations that have an interest in space applications. It essentially gathers and publishes information about global space activities. Its small Secretariat also conducts training sessions and symposia around the world. Despite its lack of regulatory powers, it can certainly be as a useful sounding board for concerns with regard to commercial space practices. It could be a key forum if it is decided that a new regulatory vehicle or convention is needed to control dangerous or harmful practices involving launching vehicles into outer space or space operations – whether these involve launch ascent or descent, in-orbit operations or extra-terrestrial issues.

Will Europe Follow the US?

The European Aviation Safety Authority (EASA) is well positioned to play a key role with regard to the emerging space plane and space tourism industry in Europe. It is already a partner in the Enterprise project led by EADS that is currently underway to develop a reliable new space plane that will operate from a European spaceport. The key regulatory question is whether Europe will follow the US example of giving the EASA the responsibility for licensing space planes, space tourism businesses and spaceports and thus controlling safety issues and concerns with regard to European space tourism operations.

Project Enterprise may therefore play a role in defining how the European space tourism business will be controlled, licensed and overseen from a safety perspective.

There will be further questions as to whether national entities will play a shared regulatory role with EASA or whether EASA will have a clearly defined transnational responsibility for European space plane activities. There will also be the issue of how EASA will interface with the International Civil Aviation Organization (ICAO) in this area if it does indeed become the responsible global entity within the UN system to regulate space tourism safety and to set international standards.[10]

Is the ICAO the Best Way Forward for International Regulation?

The bottom line is that there are actually few international treaties or international agreements in effect to provide a solid legal framework for future space exploration. Efforts to create such rules and regulations often came afoul of Cold War differences between East and West. Even after the Cold War tension had subsided, there have been many who have sought to keep a very open and unregulated approach to future space activities. Some would say that a clear legal framework that established the "rules of the road" – or the rules of the "space way" – and sets forth clearly legal liabilities, standards and regulatory guidelines actually would help the development of private enterprise in space. Others, especially in the US say let's take a *laissez faire* approach. This would be to let private enterprise develop capabilities under national regulations and then coordinate international standards and practices as needed.

Some of those who have studied the issue have suggested that an international entity such as ICAO could and should help in setting safety standards for commercial space launch systems as well as international spaceports and even ocean and atmosphere launch systems. ICAO might also be tasked to conduct "due diligence" in order to protect the public good and avoid damage to life and property where international liability might be at issue. Also, ICAO might assist in establishing explicit guidelines such as for establishing legal liability or creating a global registration process that would cover all private launch operations with a "country of record" agreeing to provide some oversight or regulatory functions related to public and air safety, frequency assignment, etc. This type of responsibility would require coordination to ensure that the ICAO and ITU did not overlap or conflict in carrying out these functions.

Most space plane advocates in the US disagree. They argue that private space missions, space exploration and space development represent a new high frontier where innovation and experimentation should be actively encouraged and that national regulatory controls are all that are needed. In light of the literally hundreds of new ideas and concepts that are only beginning to realize their potential, such efforts to set the international rules for such effort at this time would be premature.

Despite these concerns about limiting ingenuity, there would seem to be some value in chartering the ICAO to at least minimally coordinate and structure international activity related to Low Earth Orbit and proto space. An "intermediate position" might be something like an International Space Safety Board, operating on a commercial basis, under the auspices of the IAASS. This approach would presumably accomplish the needed minimal international coordination and leave to national governments the bulk of regulatory oversight.

During the early days of space flight, and continuing on until today, the ITU has registered radio frequencies associated with the launch and operation of satellites or manned missions, and continues to do so. Each national government is charged with the responsibility for the safe operation of its launch sites.

In the twenty-first century, however, we are seeing an increasingly new and different space and proto-space environment where international regulations, liability provisions and safety controls seem more and more necessary. There are spaceport projects all over the world. There could be launches from balloons, ocean based launch sites and cargo planes at almost any location. We also are seeing projects to launch people to low earth orbits via lighter-than-air craft and by the use of ion engines that will spiral out to low earth orbit over prolonged periods of time, with an increase in the possibility of being hit by space debris and micrometeorites. Further, there are no internationally recognized standards for radiation exposure and other hazards associated with a prolonged stay in space. A study of how ICAO might assume some or all of these international regulatory duties has been carried out by the McGill Air and Space Law Institute, Booz Allen and Hamilton and members of the International Institute of Space Law. The study is available on the web site of the IAASS (See www.iaass.org). At this time, safety regulators at

NASA and the FAA have shown little enthusiasm for bringing what are often seen as high-cost and bureaucratic ICAO regulators into the business of space plane regulation.

International Regulatory Controls and By Whom?

At this stage it is too early to conclude what international regulatory action needs to be taken with regard to commercial space ventures. Nevertheless, action at the international level probably needs to be taken, and sooner than later. Already unscrupulous business enterprises have undertaken to "sell" tiny particles on the Moon and Mars with so-called deeds of ownership. "Flag of convenience" spaceports might offer launch or landing facilities that might prove unsafe in terms of interference with private, commercial, governmental or military aircraft or might contribute to space debris in low earth orbit.

For these and other reasons, including providing for the safety of future "space tourists" with some form of international regulatory controls and rules of legal liability, action needs to be taken. To date, expanding the charter of ICAO to provide for the coordination of international activities in these areas may represent the most logical and most forthright way to proceed. Such action to widen ICAO's "aerospace role" would only be a preliminary step. The first step would leave open for the time being whether national commercial space transportation safety standards would continue to apply to the regulation and licensing of space planes, or whether some sort of non-profit organization entity might in the future help establish global standards for safety regulation. Such an entity would operate independently from ICAO but nevertheless in close coordination with it. Another critical question would be whether an international entity might assume regulatory control with regard to the protection of the ozone layer and other environmental concerns. So far, the ESA-backed A2 craft that would use hydrogen fuel is the only response to such environmental concerns that would seem to offer a reliable way to diminish greenhouse gases from a growing number of space plane vehicles.

The efforts to create a sub-orbital flight space tourism business is generally seen as being the first phase of a wide range of new space businesses. These enterprises may include flights to low earth orbit, space walks, stays in private space habitats and dark sky stations maintained at altitude by lighter than air craft. Beyond these ventures, there could also be commercial flights to and from the International Space Station under contract to NASA as well as private missions to support space sciences, materials research, manufacturing for the pharmaceutical industry and even support to a solar power satellite industry. In the longer term, commercial space plane technology might be replaced by other ways to access space such as through the use of nuclear propulsion, tether cables that act as a "slingshot" to lift payloads to higher orbit and even very advanced concepts such as the construction of a space elevator system.

This new and now surging development of a space commercialisation industry clearly implies the need for a regulatory process to provide for the safety of crew and passengers as well as to protect people who work at spaceports or those residents and business people who live in proximity to the launch and landing facilities.

In the US, the complete legislative authority governing regulatory oversight of space tourism is spelled out under Title 49 of the US Code, Subtitle IX, Chapter 701. US Presidential Executive Order 12465 further augmented this legislation. Then came the Final Rulemaking by the FAA, adopted on December 15, 2006 and entitled "Human Space Flight Requirements for Crew and Space Flight Participants." This FAA Rulemaking and regulatory framework establishes health requirements for space plane crews, sets safety standards for the operation of said space planes as well as for the safe operation of spaceports that protect not only the passengers but also of those that might live or work in proximity to spaceports. In the U.S regulatory framework, all passengers are

Figure 7.3: The Canadian Arrow
Astronaut Pilot Training Facility
(Courtesy of Canadian Arrow)

required to sign waivers with regard to any US Government liability and the onus is largely placed on the passengers to recognize that they are embarked on an activity that entails sizable risk.

Certainly standards for space plane operations as well as for the training of pilots are important issues that are still evolving. There are clearly few international facilities available for this purpose since US and Russian manned space programs have dominated human exploration programs to date. In Canada, the go-ahead Canadian Arrow and Planetspace organization has established a sophisticated training facility for space tourism pilots. But so far, the standards to be used for such training have not been published.

Other Dangers for Space Tourists

Despite constraints imposed by international due diligence procedures in force through the ITU and the UN COPUOS-backed IADC, the increase in orbital debris has continued to escalate. Although the IADC has finally come up with seven "Guidelines" to address the problem, these guidelines, based on international consensus, are vague and contain a number of serious loopholes in terms of implementation. Perhaps more significantly they are not backed by any explicit incentives or sanctions.

The recent Anti Satellite (ASAT) missile firing by the People's Republic of China that destroyed an aging Chinese meteorological satellite and created hundreds of new debris elements, followed by the breaking apart of several Russian satellites, has only accentuated the problem. International treaties, regulatory procedures and technical solutions for removable of such debris currently remain elusive. Some experts suggest that we are reaching a "critical point" where the breaking apart of a number of satellites will create a deadly avalanche effect that lead to low earth orbits becoming increasingly deadly.

Another major concern with regard to the safety of space tourists and the further development of the space tourism industry is the deployment of weapons – either offensive or defensive – in outer space or on the Moon. Strict international treaties and conventions that prohibit the deployment of weapons or anti-satellite devices are seen by some as key elements of a longer-term strategy for not only the development of space tourism but successful human travel to the Moon and colonization of space assets.

The new US space policy announced by Executive Order in August 2006 suggests that the US will not be bound by existing space-related treaties if it feels its national security is threatened by attacks on critical space assets. These issues are addressed in more detail later in the chapter about strategic and military issues.

The Next Safety Related Steps

The FAA rules now appear to be a more or less acceptable basis for proceeding with the development of space plane systems and for the initial operation of spaceports and space tourism business. For better or worse, these are the regulations that will apply for the first few years since the first flights seem destined to come from US spaceports and by US developed and manufactured space planes. Projects from Europe, Canada and elsewhere are now scheduled to follow suit. As these international programs emerge, US precedents will undoubtedly be closely observed by others and most likely followed unless major accidents suggest alternative courses of action.

Clearly, as experience is gained in the US and abroad, these regulatory processes and licensing operations will need to be adjusted and perfected. It is our belief that the "perfection process" could and should be started now. Thus it is recommended that the FAA, in cooperation with NASA, should be directed to work with a White House Commission on Personal Spaceflight Travel Safety and Security that would be a direct parallel to the previously successful White House Commission on Aviation Safety and Security.

This Commission would be directed to complete its work by 2010 and be given a specific charge to carry out the following program:

- Devise ways to reduce personal space flight accidents and enhance flight safety by developing targeted and realistic objectives;
- Develop a charge for the FAA (and NASA as appropriate) in terms of creating standards for continuous safety improvement, and these goals should create targets for its regulatory resources based on performance against those standards;
- Develop improved and more vigorous standards for certification and licensing of space planes and their operations, spaceports, and training and simulation facilities;
- The Federal Aviation Rules for Private Space flight should by 2010 be rewritten with statements in the form of performance-based regulations wherever possible;

- The FAA should develop better quantitative models and analytic techniques to assess space plane (and especially space plane subsystem) performance and to monitor safety enhancement processes. These should be based on best industry practices as new operational vehicles and safety and emergency escape systems come on line;
- The FAA Office of Commercial Space and the Department of Justice should work together to ensure that full protections are in place, including new legislation if required, so that employees of the space tourism business, including but not limited to manufacturers and/or operators of space planes, maintenance and other ground based crew, owners and operators of spaceports and owners and operators of personal space flight training and simulator facilities, can report safety infractions or risk factors of concern regarding safety violations or security infractions to government officials without fear of retaliation or loss of employment – i.e. full "whistle-blower" protection for such employees. This would include the creation of a safety hotline that is similar in operation and purpose as the current NASA Safety Reporting System.
- This effort should use the White House Commission on Aviation Safety and Security as a model for this activity.[11]

The above program would deal with public and passenger safety requirements in the US, but in order for the space tourism business to flourish safely on a worldwide basis, other space-faring nations will need to develop parallel measures – and a degree of international coordination and cooperation will also be an essential ingredient. For the time being, bi-lateral coordination among countries may be sufficient, but at some time after 2012, some of the issues regarding the role of ICAO in space will need to be definitively answered.

References:

1. *Cosmic Log* – Tuesday, October 24 2006 – by Alan Boyle
2. *Wall Street Journal* – "How Safe Is the Race To Send Tourists into Space?" – *April 19, 2007*
3. Space Tourism Set to Lift Off – *The Age.co.au* – 29 May 2007.
4. Ken Alltucker – *The Arizona Republic* – May. 16, 2007
5. Space Travel.com – http://www.space-travel.com/reports/Wyle_
6. FAA, *"2006 Commercial Space Transportation Developments and Concepts: Vehicles, Technologies and Spaceports,"* Washington, D.C. January 2006
7. Joseph Pelton with Eric Novotny – NASA's Space Safety Program: A Comparative Assessment of Its Processes, Strengths and Weaknesses, and Overall Performance, (September, 2006), Washington, D.C. www.spacesafety.org
8. International Association for the Advancement of Space Safety – www.iaass.org and IAASS Report, August 2006
9. International Telecommunications Union – www.itu.int
10. European Aviation Safety Agency web site – http://www.easa.eu.int/home/
11. *White House Commission on Aviation Safety and Security: The DOT Status Report – http://www.dot.gov/affairs/whcsec1.htm*

— Chapter 8 —

"If we could be really clever, couldn't we design reasonably priced, clean and efficient hypersonic transportation systems and space tourism vehicles that ran on hydrogen and would not pollute the stratosphere?" Joseph N. Pelton, "The Next Billion Years"

THE POSSIBLE SHOW STOPPERS?

Sir Richard Branson's high flying parties on his private island and his beaming, and ruggedly handsome face on the cover of *Time* have stimulated public interest and fired the imagination of potential space tourists around the globe. His well-tuned publicity machine has kept private space flight in the headlines, including the announcement in April 2008 of the first space wedding, scheduled to take place on one of the first Virgin Galactic flights.

He even partnered with Google to announce the launch of a new space exploration enterprise called "Virgle" which captured some headlines before it was recognized as an elaborate 'April Fool' hoax on April 1st 2008!

Charles Simonyi's 13-day "First Nerd in Space" trip to the International Space Station in April 2007 was a tremendous PR success – both for him and the space tourism industry. Simonyi's girlfriend, the American TV personality Martha Stewart, certainly helped extend the coverage. Her exotic lunch bags packed with duck pate for her boyfriend in space gave the mission a romantic twist.

Increasingly, the enthusiasm for space tourism has certainly commanded a lot of coverage in the world-wide media. Even gossip columnists are poised for the day when a superstar like Madonna or Bill Gates joins the celebrity rush to zero G. And on the horizon are the juicy news prospects of trips to private space hotels and even a fare-paying journey around the Moon.

Meanwhile, the various ways that private citizens can now go into space continue to expand apace and the weekly total of those taking options on space plane flights has not slowed; indeed it's been fueled by the publicity. This included Jeff Greason's shrewd move in 2008 to offer a "budget flight" to an altitude of some 60 kilometers (i.e. 37 miles high) on board his new 2-seater Lynx space plane, which may very well attract some stars whose fame may be waning just a tad.

Space Adventures has demonstrated wily resourcefulness in arranging an ever-expanding array of alternatives. Billionaires, millionaires and the merely well-off can now find some sort of space adventure to enjoy. These possibilities begin with a ride on the so-called "vomit comet" for about $4,000, where you can soar at the top of a parabolic arc and experience almost a minute of weightlessness. Indeed you can get nearly four minutes of "hang time" if you sign up for four parabolas, which is what Professor Stephen Hawking eventually got on his famous "ride to space." Those with a slightly thicker wallet can take to a simulated ride to space for around $8,000, now available at Star City near Moscow. Then there is a ride in a Russian Foxbat plane that soars to the dark sky near the edges of space for an even larger pile of cash of just under $20,000. The options then continue up to the very high-end trip to the space station for the current price, now increased to around $35-million. Nor is this the end of the spectrum. Space Adventures has begun to explore the future possibility of space walks and has announced a privately-financed trip to orbit around the Moon for a cool $100-million price tag.

Literally hundreds of rock stars, sports legends, movie stars, corporate moguls and now a British royal princess are among those who have already signed up for sub-orbital flights with Virgin Galactic and its competitors. But does that make space tourism a serious business that can truly achieve market viability and respond to the many challenges and issues yet to be resolved?

As we've already learned, there are certainly serious market studies by serious organizations such the European Space Agency, the Futron Corporation and others suggesting that space tourism has the potential to become, at least in time, a "real" multi-billion business. The size of potential markets has been probed, options critiqued, and sensitivities related to price, risk, and other variables have been tested.

These market studies have certainly led to questions as to whether a string of new offerings may be needed to sustain an on-going space tourism industry? The capital investments made by entrepreneurs and by state and federal departments now total over a billion dollars. These serious investors and government officials have signed off on a range of investments that include space plane developments, new spaceports and training and testing facilities. The biggest market of all could be projects like the U.K.-led A2 hypersonic hydrogen-fueled vehicle that could eventually fly 300 high-flying passengers from Europe to Australia in under five hours.

The Space & Advanced Communications Research Institute at George Washington University released an in-depth report in April 2007 entitled "Space Planes and Space Tourism" (available on line at www.spacesafety.org.) It provides information about the various global efforts to develop space planes, lighter-than-air craft with ion engine systems, "dark sky" stations, tether-based systems, and space elevators. There is an amazing array of projects out there to get people off the planet and ultimately some of these are likely to succeed. And it doesn't end there. As detailed earlier, Bigelow Aerospace of Las Vegas and Inter Orbital Systems (IOS) of Mojave, California are among those seriously embarked on deploying space tourism facilities in orbit – and as soon as 2010 to 2012, or so.

Also, there are now efforts by various governmental and commercial bodies to develop serious projects related to space tourism in Argentina, Australia, Brazil, Canada, China, France, Germany, Malaysia, Rumania, Russia, Singapore, Switzerland, Ukraine, the United Arab Emirates, the United Kingdom as well as the United States. These various indicators suggest a dynamic new industry and a nascent market that is ripe for successful development and potentially explosive growth – figuratively and literally.

But amid all the enthusiasm and media hype, there are a number of real world issues to address and very specific problems still to be solved. Not least, this type of enterprise is a "high risk" undertaking, in several senses of the word. The George Washington University study, undertaken in 2006 and 2007, indicated that the "churn rate" of corporations involved in space tourism will likely remain high. For every space tourism company that succeeds there may be two or more that fail.

It is against the background of this fast-changing industry that the team at the FAA Office of Commercial Space Transportation has striven hard to develop safety regulations that will protect the US Government and property adjacent to spaceports. People like Patricia Grace Smith, George Neild, and Kenneth Wong have been clearly devoted to developing a new US space plane industry and making it safe – especially in terms of public safety. FAA Office of Space Commercialization regulations also provide reasonable safeguards for crew and passengers, and provide a regulatory framework that can allow this industry to survive and even thrive. But regulatory provisions are not the only things that the space tourism business needs to succeed. There are, in fact, several key questions still to be addressed. These include what might be called some potential "show stoppers":

- **Environmental Concerns and Issues:** As described in a previous chapter, the flights of the supersonic Concorde into the high stratosphere were a serious concern in terms of its potential damage to the ozone layer. Many breathed easier when the SST was grounded. The prospect of potentially thousands of flights by space planes into stratosphere raises anew these environmental concerns. Likewise the near-term development of supersonic commercial executive jets as a parallel industry raises similar questions with even greater concern. The truth is that damage to the ozone layer may be a more urgent concern than global warming. Genetic damage could kill off the human race much faster than rising temperatures. This may seem like a quibble to some, but survival of the species seems deserving of some serious thought. The FAA currently has overall regulatory responsibility for the space tourism industry and Associate Administrator Patricia Smith before leaving office at the end of 2007 explained at a public forum sponsored by the Center for Strategic and International Studies (CSIS) that they do have environmental engineers examining this issue. Yet before full-scale service begins it would seem that the Environmental Protection Agency (EPA) which has ultimate jurisdiction in this area under the National Environmental Protection Act (NEPA) should examine in some depth the potential environmental impact of a large number of flights into the stratosphere and near-space by space planes and supersonic commercial executive jets. This should be an examination in terms of both the ozone layer and other green house pollutants. Further, this is much more than an American concern. Environmental scientists around the world should examine this issue and render an opinion on both the threat level and even more importantly possible solutions. What we do know is that most of the space planes now under test and development spew out either greenhouse gases such as carbon oxides or even worse (some 400 times worse according to some ecologists) nitrous oxides. Perhaps even more important we know that new technology such as hydrogen fueled engines would produce much more benign water vapor as its residue.

- **Risk Management for a Space Tourism Industry:** The first words in the waiver statements that "citizen astronauts" will be asked to sign indicate that such flights are of high risk and that the spacecraft are "years away" from being certified as "safe" by the FAA. It is not clear whether there is a reliable insurance or reinsurance market that will be able to sustain the space tourism industry on an on-going basis. There is an even bigger question as to whether the various start-up companies could possible provide self-coverage. This might possibly work on a short term basis but certainly not for the longer term. So far third party liability coverage by insurance companies has been elusive. This is in part, because insurance companies are leery of the fact that the majority of passengers signed up for flights are fat cats with lawyers at their command. Of the 250 passengers who have posted some $35-million in deposits with Virgin Galactic, virtually all are high net worth individuals who families would command the resources needed to hire high profile lawyers that might seek to overturn signed waivers of liability by the passengers if a fatal accident occurred. This would be especially true if the operator or the FAA oversight seemed demonstrably at fault.

- **Other Regulatory Concerns:** Even after the waivers and insurance questions are resolved there are a number of other regulatory concerns awaiting the space tourism companies almost at every turn. These companies, at least those in the US, need to make sure that even if they have jumped through the right hoops at the national level that State laws or local ordinances might hold them responsible for large claims if there should be an accident involving public safety on the ground or claims by the families if there is an accident on a flight. In Europe there are other issues that seem rather prosaic or just a detail, but nevertheless involve a large price tag. In Europe for instance Value Added Taxes (the dreaded VAT) often run to 25%. Thus a $200,000 trip could end up costing the passenger $250,000. In Sweden, where the Esrange airport is being converted for Virgin Galactic passengers wanting to see the Aurora Borealis up close and personal, relief legislation is being considered. This would allow space tourism flights to be taxed in the same way as hot air balloon flights, allowing passengers to pay far less than the current 25% VAT.

The current trend in Europe is to give the European Aviation Safety Authority (EASA) much more control of all "aviation" flights in the region and for national aviation regulatory systems to surrender their authority to this process. This would serve to lessen the ability of a country seeking to provide regulatory, tax or liability relief to do so if the European Union and the EASA has assumed overall control. The space tourism industry will presumably have to address and resolve such issues in each legal and taxation regime around the world.

Finally in the US those seeking to develop new space plane technology are today significantly constrained by the International Trade in Arms Regulations (ITAR) that limit the export of space plane-related technology that can be and will be increasingly available in Europe, China, Russia and other parts of the world. This issue, as discussed below, is an anchor that will be a drag on advanced technology in the US In time, this will allow other countries to catch up with American innovators in the field. US firms may currently lead – but not for long if ITAR restrictions are not relaxed.

- **Weaponization of Space:** The August 2006 National Directive on Space from the White House indicated support for private initiatives in space and space commercialization. The same directive also indicated that the US would take all necessary actions to protect space assets and otherwise implied that further "weaponization" of space could occur. It would appear highly desirable for a new and clarified National Directive to be introduced that addresses more clearly the regulatory and strategic position of the US Government with regard to space tourism, private space stations, the deployment of military systems in space, etc. Even more desirable might be an international convention or treaty that clarified such issues on a global basis among all space faring nations.

- **Orbital Debris:** The increase in orbital debris, particularly in low earth orbit, continues to expand despite the due diligence requirements developed by the ITU. The Inter-Agency Space Debris Coordination Committee (IADC) acting under the auspices of the UN and COPUOS requires the space agencies to consult together on this issue.

Indeed the growing problem with orbital debris is taking on scary proportions, particularly since the Chinese anti-satellite (ASAT) test in January 2007 that destroyed the Fengyun 1C meteorological satellite led to the largest single source of orbital debris since the start of the space era. This missile test generated some 2000 debris elements exceeding 5 cm in size, which is large enough to destroy a Space Shuttle window or significantly damage the International Space Station. The solar wing on the Hubble Telescope was severely damaged by a piece of orbital debris. As the number of significant pieces of space junk mounts into the tens of thousands, the effective use of space for peaceful, commercial or strategic purposes diminishes, seemingly every day.

Then, in February 2007 there were four other new sources of major space debris. The break-up of two Chinese satellites and of two Russian large launch vehicle components added hundreds more debris elements. Corrective action could have prevented the spread of much of this debris.

A year later, in February 2008, it was reported that the US spy satellite known as satellite 193 was destroyed in orbit by a missile fired from a warship in the Pacific Ocean. Reports indicated that 99% of the debris from this satellite de-orbited within days of the missile firing. This was, in part, due the fact that the satellite was already in de-orbiting mode.

However, such a violent solution is not really practical. Equipping all satellites with fuel to either de-orbit or move beyond the GEO orbit is actually the ITU mandated solution. Venting of fuel so that a satellite does not explode is another critical step. The line is that orbital debris, despite efforts taken to date, is a mounting problem, especially for low earth orbit. Fortunately orbital debris is much less of a problem for sub-orbital flights but for those planning trips to space stations, the risk is very real. Just a mere chip of paint a few millimeters in size is enough to crash through a space shuttle window and cause a fatal accident.

The UN Committee for the Peaceful Uses of Outer Space (COPUOS) has had this problem under "study" since 1994, but the problem has only grown worse in the last 15 years. The Inter-Agency Space Debris Coordination (IADC) Committee has finally come up with seven "Guidelines" to address the problem, but these guidelines, based on international consensus, are vague, riddled with loop holes and legally non-binding. The ultimate adoption of these guidelines on a unanimous basis within the UN will, at best, represent a baby step of progress on orbital debris.

Meanwhile hundreds of billions of dollars of space assets from remote sensing and telecommunications to spy satellites are increasingly at risk. GPS navigation satellites, among others, that are used not only to target missile firings but also to allow planes to take off and land all over the globe could easily be damaged by growing amounts of space junk in orbit – especially low earth orbit. From the Internet to global electronic funds transfer, from close satellite monitoring of hurricanes and tornadoes to our defense and civilian space assets, we live in a modern technological society that is more and more vulnerable if our satellites in the sky should fail. A new effort to address this growing menace to space flight, space exploration and space applications is clearly needed – not just for space tourism, but for all space-related systems and missions.

Clarification of US Governmental Roles in Space

The law under which the FAA now regulates space tourism indicates essentially two things. One is that the FAA is directly responsible for the public safety of the surrounding populace and proximate air space as well as that of those involved space plane operations. This is their prime duty. Passengers and pilots and crew for space planes fly at their own risk with the knowledge that what they do is very high risk. The other current responsibility of the FAA Office of Space Commercialization (FAA-AST) is to promote the growth and development of this new industry. The FAA-AST has worked closely and effectively with the Private Spaceflight Federation to develop what appear to be reasonable safety regulations, and the rest of the world seems to be following similar approaches in response to the US lead. Putting agencies in the role of both "cheerleader" and "referee," however, is always a flawed concept. NASA's role, serving as its own safety monitor for the Space Shuttle and the International Space Station (ISS), was clearly exposed as a problem by the Columbia Accident Investigation Board and the earlier Challenger investigations. The nuclear industry has its regulatory controls handled by a separate government agency, but it is not charged with developing nuclear power as well. For a number of years the FAA was charged with promoting civil aviation, but legislation was enacted to remove this responsibility after a major airline crash over a decade ago. A review of all these issues, such as whether the Department of Commerce rather than the FAA-AST should promote space tourism, should be reconsidered after regularly licensed flight begin to take place after the experimental period is finished in 2012. At the same time consideration should be given to the possibility that aerospace research might be transferred from NASA to another governmental agency. The progress being made by the European Space Agency to develop A2 hydrogen powered hypersonic craft that could address global warming and safety issues is a case in point. Currently languishing NASA and FAA aeronautical research programs suggest that there are serious policy issues that Congress needs to consider as the space tourism industry moves forward.

Environmental Concerns

Most aircraft emit a good deal of pollution and indeed it has been estimated that 15% of the pollution that contributes to global warming may come from aircraft. One of the reasons why the Concorde SST generated so much concern was the fact that it flew very close to the ozone layer, where destructive chemical reactions are more pronounced.

The Concorde, created as a joint venture between the British and French nearly 50 years ago, was designed to travel at a speed near Mach 2 for transatlantic flights at the high end of its trajectory. The main concern for environmental groups at that time was the effect of the plane's engines on the ozone layer that resides at the top of the stratosphere. Further the exhaust from the Concorde was seen as compounding the air pollution problem. The plane was designed to climb to an altitude of 19 kilometers (or 60,000 feet), nearly 7000 meters higher than "normal" airplanes. This high altitude for the SST was key to achieving very high velocity, but it also put the plane proximate to the ozone layer. The engine also required an exotic fuel that when burned, emitted nitrogen oxides (NOx) directly into the ozone layer. Nitrogen Oxides are considered to be up to 400 times worse as a greenhouse gas than carbon dioxide.

Another exhaust problem is the emission of water vapor and hydroxyl radicals (OH). Each of these types of exhaust emissions contributes considerably to the greenhouse effect. This means that even a hydrogen fueled hypersonic jet, may not be a magic solution to high altitude pollution by high performance jets.

The greenhouse effect is simply the warming of the lower atmosphere, allowing the upper level of the atmosphere, the stratosphere, to cool. Even minor changes in temperature in the atmosphere have a dramatic impact on weather patterns across the globe. In particular, ozone depletion from jet emissions allows more ultraviolet radiation into the lower atmosphere. This UV radiation at high enough levels leads to genetic mutation in all animals including humans. One should trust us when we say genetic mutation is a real "no no" when it comes to preserving the human race.

The other thrust of the anti-Concorde campaign was to prevent the United States from expending scarce national funds into the creation of an expensive airplane with limited passenger capability, both in terms of range and number of people that could be accommodated in a single flight. Again there are parallels here in that space tourism planes and commercial supersonic executive jets could be subjected to similar criticism and environmental complaint at a time when global warming concerns are peaking.

Environmentally, it was estimated in the 1970s that a large fleet of supersonic aircraft in constant operation would reduce the ozone layer faster than the total worldwide output of chloro-flouro-carbons (CFCs). These CFCs, in terms of length of chemical interactions, are today still considered to be among the largest threats to the ozone layer. Interestingly enough, the debate in the 1970s centered exclusively on the Concorde, whereas the US – and other countries – also had, and still have, a fleet of supersonic bombers and fighters that produced the same emissions at similar altitudes.[1]

In the 1970s and 1980s the issue of high altitude pollution were of major concern, but for obvious reasons linked to global warming are even greater today.

Today there are proposals to operate not only a number of fleets of space planes to support space tourism, but also to create supersonic (Mach 2) commercial executive jets to ferry CEOs around the planet at ever faster speeds. The environmental impacts of these new jets – both for flights to "nowhere" and to "somewhere"– are not clearly understood. It would appear extremely important to examine the possible dangers in terms of global warming, depletion of the ozone layer in terms of UV radiation in the context of genetic mutation and other possible adverse environmental effects before these commercial systems begin operation. It would seem very important to ask a basic question. This would be: "If we could be really clever, might we not design reasonably priced, clean and efficient space tourism systems or hypersonic executive transportation vehicles that would not pollute the stratosphere?" Such systems might be propelled by entirely new systems. These vehicles might employ magnetic levitation (Maglev) technology. New types of lighter-than-air craft could be augmented by ion engines. Nuclear engines, space tether systems or other technologies might be developed. Such future "space systems" should be seriously considered, not only in terms of cost, efficiency and environmental effects but also in terms of long-term safety.

For centuries, global industrial enterprise has been driven by the development of new technology that allows more and more rapid economic throughput and "industrial and GNP growth." It turns out that the 21st century may be the time that the main driver of technology and new systems could turn toward survival of the species and cultural and social objectives such as improved health. When humans have achieved a high level of material wealth, it may just occur to them that the survival of *homo sapiens* and our great, great grandchildren may just be more important than a larger bank account. If the planet were wiped out by a comet or meteorite or Near-Earth Object, or we experience massive tidal flooding due to global warming, economic growth models may well change. Likewise if the hole in the ozone layer continues to expand and there is large scale genetic mutation whereby newborn babies start to grow three arms and damaged brains, then the acquisition of a shiny new penthouse condo or a Bentley may suddenly seem a whole lot less important. In such an environment we may start to think that developing technologies and economic systems geared to saving the species might be a good thing to do.

This is not to suggest that there are not solutions to these environmental issues and concerns that might also spur economic prosperity. It may be that with proper controls and contributions to mitigation processes, the threat of space planes to the environment can be controlled and minimized. We are just mounting a call for "smart" strategic planning now. There could well be "win-win" solutions that allow us to develop better transport systems and to fly into space and still be ecologically sound.

We believe that in the US, the Government – and the EPA in particular – should be examining the environmental issues and working with engineers and scientists to devise ways to limit adverse environmental effects to much more acceptable levels. Most importantly of all, we suggest that this is something to do now – up front–before the damage occurs. Too often, when it comes to issues like global warming and ozone layer depletion, we end up entering the fourth quarter three touchdowns behind … or is it starting the ninth inning six runs down … or is it seeing the clock about to expire and needing three goals just to tie. Regardless of which tired sports analogy we use, we need to recognize environmental problems sooner and stop playing from way behind.

It may well be that smarter, greener and truly innovative technology may provide us with new answers for cleaner and safer space systems. Recovery from all sorts of polluting practices with smarter systems and much smarter and cleaner technology may well be the best hope not only for human survival, but actually also for new economic growth. New environmental technologies might trigger a new wave of industrial development and economic growth. Once this is understood, big business might even learn to embrace the tree huggers.

One can at least hope that innovations in "smart" transport, clean energy production, and pollution-free info-systems may drive an economic revolution that might sustain productive human enterprise for, perhaps, another hundred years. Space opportunities as part of "green systems" may well become a part of this new economic opportunity. Environmental futures and green systems, if planned and executed cleverly, may offer employment, prosperity, and a smarter path to space-related industry and even off-planet colonization. Right now, however, these environmental issues deserve urgent attention. EPA and environmental authorities around the world – it is time to get on the ball.

Risk Management for a Space Tourism Industry
The challenges to the space tourism industry go beyond environmental concerns. The space tourism industry, even more than other space applications needs a robust and well informed risk management system that can help these new enterprises cope with the many challenges – and the potential huge losses–that it faces. Risk issues become even more vital as a new and incipient technology such as space tourism takes off. It is believed that as space planes mature and are licensed by the FAA the number of passengers choosing to fly will increase. Prices will drop, launch operations will become smoother, easier and cleaner, and economies of scale will take over to help stabilize the industry.

The problem is how does the industry bridge the gap from high-risk start-up to on-going reliable and safe operations? Courtney Stadd, the well-seasoned former official at both the US Department of Transportation and NASA who has some three decades of experience in commercial space efforts, has had this to say about the prospects for space tourism:

> "Space entrepreneurs still tend to be seduced by the tendency to mistake technical possibility for market opportunity. Nonetheless, I feel that, by and large, today's entrepreneurs represent a particularly sophisticated and seasoned group of business managers who stand a better chance of navigating the many daunting technical and market challenges associated with the new commercial space industry sector. Although it is still a work in progress, the good news is that the US government is doing a better job of fostering a more stable and predictable regulatory and policy climate for space entrepreneurs."[2]

Despite Stadd's optimism, the fact remains that many of the space tourism businesses out there today are busily designing and testing space planes with a focus on the technology. It is the "gee whiz" technology that generates the adrenaline rush that comes from opening up a new frontier. As we know from long experience, pioneers – both territorial and technological–often end up dead on arrival from "bleeding edge" technologies.

Most of the communications satellite ventures that today rake in billions of dollars in revenues and large corporate profits would never have made it through the early development years unless they had been able to purchase launch insurance against the sad event that their satellite and launcher failed during lift off. Lloyds of London and some innovative brokers of the 1960s and 1970s were able to convince large insurance companies and re-insurance alliances to spread the risk of such launches broadly. Over time they would be able to make money from this new industry by playing the odds that a few launches might fail, but that by due diligence and proper calculation of the risk a business could be built on "managed risk."

The problem with the space tourism business is that many consider it too early in the development curve to provide reasonable levels of coverage at reasonable rates. In the 1960s when the first coverage began for the satellite industry, it involved insurance against the potential loss of the satellite, the loss of the launcher, and even the very remote possibility that the rocket might go awry and land in Miami Beach or some other populated area before the destruct button could save the day. (Actually, in the early years, the wary insurance companies were only willing to provide Intelsat, then a monopoly global consortium, with insurance against the second launch failure in a row.)

The BIG difference for space tourism is that there are people on board. Liability claims for people and their lives are something else. The FAA regulations have attempted to address that issue with a rule-making process whereby each and every passenger engaged in a space tourism flight would receive a detailed briefing on the craft in which they are planning to ride plus a detailed review of the possible high level of risks that would be involved. Then most space tourism companies would proceed to require their passengers to sign an iron-tight waiver of all legal rights in light of the disclosures made to them of the risks involved. Thus there is one waiver that holds the US Government completely harmless in all regards against any loss of life or damages. Then the space tourism companies ask their customers to sign a parallel waiver. The first five citizen astronauts flew with the Russian Soyuz under arrangements made by Space Adventures signed a similar release for their flights.[3]

It is too early to know whether the global risk management companies, the insurance and re-insurance industries and others will be willing and able to provide the extra layers of coverage that the various space tourism companies need to become viable. If there are no accidents and several years of positive experience are achieved before the first space tourism mishap materializes – especially if there are perhaps injuries, but no fatalities – the viability of the overall enterprise will become clear. The time period between 2009 and 2012 will thus be critical. Some daring entrepreneurs from the global insurance industry have made possible the now vibrant space applications businesses, enabling the launch of communications and remote sensing satellites, that contribute over $100-billion a year to the world economy. It is still too early to know how today's men and women in gray suits equipped with actuarial tables and computers will respond to the new challenge.

Weaponization of Space:

Today, the aerospace industry is a several hundred billion dollar enterprise worldwide, but the launch vehicle business is only a small percentage of this total. In fact, the commercial launch business experienced a state of decline in the early 2000s with commercial launches sinking to a level of only 12 to 15 per year during this period and sometimes lower. More recently, there has been an increase in demand and the number of commercial launches has grown since 2004 to an annual level of around 20 per year. Nevertheless, the demand for military aerospace systems has outpaced all commercial space growth in recent years. The prediction made by many around the year 2000 that commercial space would outpace military space activities and represent more than twice the economic level of expenditures by 2007 simply did not occur. Military space is still the largest gorilla in the global economic jungle of outer space affairs. Warfare in Iraq and Afghanistan may well have been part of the reason. Stiff competition from fiber optic cable systems has certainly limited communications satellite growth while remote sensing by satellite has yet to achieve its potential.

The US National Security Policy Directive of August 2006 concerning American space policy was announced rather unilaterally from the White House, with a minimum of consultation with Congress or with other civilian think tanks. The pronouncement can probably best be described as "puzzling" as to its implications for the future of commercial development of space – including the future of space tourism. On one hand this statement clearly indicates that the United States is "committed to encouraging and facilitating a growing and entrepreneurial US space sector." This would seem to be a good thing for commercial space applications, including space tourism. But the statement does not stop there. The Directive goes on to say that the US Government will use the commercial space sector "to the maximum extent possible consistent with national security." Even that sounds to the good.

But as is the wont of many governmental pronouncements, the statement continues and ventures into what might be very dangerous waters – or should we say the "murky ethers of outer space." The Government (or at least the White House Directive) indicates that there is to be an increased focus on the strategic importance of outer space. This Directive sets forth at least two principles that puts the utilization of outer space in both a security and military context:

> "The United States considers space capabilities – including ground and space segments and supporting links – vital to its national interests. Consistent with this policy, the United States will: preserve its rights, capabilities, and freedom of action in space, dissuade or deter others from either impeding those rights or developing capabilities intended to do so; take those actions necessary to protect its space capabilities; respond to interference; and deny, if necessary, adversaries the use of space capabilities hostile to US national interests.

> "The United States will oppose the development of new legal regimes or other restrictions that seek to prohibit or limit US access to or use of space. Proposed arms control agreements or restrictions must not impair the rights of the United States to conduct research, development, testing and operations or other activities in space for US national interests."[4]

Oh boy! One does not know whether to sell any stock one might have in a space- related company or buy stock in a major military aerospace supplier. If one re-reads the statement again slowly, you might be inclined to do both. In short, this does not sound entirely promising for the civilian-oriented space tourism business. The sabers in space seem to be rattling. New international agreements to control space plane traffic, under the auspices of the ICAO or other entities, would seem difficult to implement if the US Space Directive is to be taken as meaning what it says.

This statement of principles, as set forth in the Directive, would seem to suggest that the US strongly views outer space first and foremost in a strategic and military context. Such a stance, if taken literally and compared to statements by the USSR. during the Cold War, would seem to make it at least questionable for commercial organizations to invest in "space deployed infrastructure" or even orbital tourism enterprises if they feared that their high capital value investments were to be subject to military attack or perhaps appropriation for military or national defense purposes.

One can also look at the Directive from yet another angle. Many of the commercial enterprises listed in Appendix A as engaged in development of space planes, or otherwise engaged in providing access to orbital systems, are being encouraged in their efforts by either NASA or the US Department of Defense. Companies such as Space X, XCOR, SpaceDev, Alliant ATK, etc. have or have had major US governmental backing. At this time there are essentially three types of enterprise that are pursuing what might be generally characterized as the "space tourism" or "commercial access to space" market –

- Major Aerospace Corporations. Represented in this category are such companies as Boeing, Lockheed Martin, EADS, Northrop Grumman, etc. These organizations are typically heavily funded by governmental agency or military contracts – sometimes both.
- Mid-Level and Major New Entrepreneurial Corporations. These organizations are well capitalized, and may also be funded by governmental, military or industrial backers. Yet these organizations are still entrepreneurial in their approach to developing new launch systems and are often noted for breakthrough innovation. These entities would include Scaled Composites, Space X, SpaceDev, etc. In many cases, very successful industrialists such as Paul Allen, Jeff Bezos, Sir Richard Branson, or Elon Musk have infused a good deal of their own capital into these organizations.
- True, Small, Start-Up Entrepreneurial Organizations. This category spans various types and sizes of organization. They include those who are proposing an entirely new way to access space, such as the "volunteer based" JP Aerospace or Space Elevator Inc. These "new ways to space" organizations consist of only a small group of people and small capitalization, but have truly innovative ideas. There are other start-ups that believe they can adapt existing rocket motors or technology to create cost-effective ways to space without having large staffs or large capitalization. Finally, there are organizations that represent little more than a flashy web site and an inspired dream of getting to space.

In many ways, today's space tourism enterprises resemble the historical patterns that were seen at the outset of the aviation industry and the first attempts to provide various types of air transport services. Or it might be compared to the early days of the telegraph industry in the United States. The FAA's current approach of undertaking a "case by case" review of each company seeking to be licensed to operate an "access to space" vehicle or to operate a spaceport thus seems appropriate at this early experimental stage of development. There is no way that general standards and specifications can be set in today's heady environment of innovation and breakthrough concepts without stifling new invention or favoring the largest organizations over the start-up enterprises. Nevertheless, there does appear to be a need to move to a more strict oversight role as technology evolves and subsystems, at least, become more proven in their performance. In recent hearings held in Washington, D.C. the Chairman of the Transportation and Infrastructure Committee in Congress indicated he felt that there is a need for a "change of attitude at the highest levels of the FAA." His remarks referred to a number of safety problems with regard to aircraft wiring, structural problems with wing structures, etc. that have occurred in the early part of 2008. These hearings had no relation to spaceplane projects, but the broadly based concern expressed in these hearings is that the FAA had once again become too closely associated with the airline industry in a way that could endanger passenger safety. This was strongly denied by FAA spokespeople who noted that a review of safety records indicated a "very, very high compliance rate." It is not until 2012 before Congress will re-examine the regulatory oversight process and indicate whether it wishes a more strict review process, but clearly performance by the FAA in other sectors will likely be a part of the review process.

Another key question is whether the mainstream of space plane and space tourism development will be largely dominated or at least heavily influenced by military programs and objectives? Already NASA has turned over its own most advanced space plane development – the X-37 and X 43 – to the Air Force.

Over 85% of the companies seeking to develop new space planes or other means to access low earth orbit are US enterprises and obviously subject to US laws and export controls regarding technology and licensing arrangements. US-based space tourism is thus easily controlled or co-opted by the military. Virgin Galactic has been struggling with access to development details of SpaceShipTwo due to restrictions imposed by the International Trade in Arms Regulations (ITAR) and its review and control of strategic information.

However, there are also spaceplane initiatives around the world outside of US arms R&D control. These include the efforts in Argentina, Brazil, Canada, China, France, Germany, Israel, Rumania, Russia, Switzerland, and the United Kingdom. These initiatives are focused on developing new launch systems; and spaceport initiatives ranging from Australia to the Middle East. China, Malaysia and Japan could soon join into the development process.

In the US, legislative and regulatory efforts to control these entrepreneurial initiatives from a public and passenger safety viewpoint are now well established. The FAA has set health standards, specified safety controls for launch operations and also aided in the process to establish insurance and risk management. Likewise the FAA has, on a case-by-case basis, undertaken to license companies and their launch operations so as to provide safe operations and set controls to limit possible damage to public and private property. These regulatory actions will in many ways set precedents to be followed in other countries. What is not clear is the degree to which the US military might seek to exert policy or even actual control over the emerging new space tourism business. At this stage the White House National Security Policy Directive of August 31, 2006 seems to raise more questions than it answers.[5]

Orbital Debris: Issues, Concerns and Potential Solutions

Orbital debris generally refers to material that is in orbit as the result of space missions, but is no longer serving any function. There are many sources of such debris. In addition, there are also micrometeorites, so-called meteor showers (such as the Leonids), and other natural sources of radiation including the solar wind and very powerful solar storms. These sources of natural debris, cosmic radiation and other phenomena that could harm spacecraft or endanger astronauts for the most part remain fairly constant and the magnitude of their danger can be reasonably calibrated or special alerts issued. Debris from various space programs and the aftermath of anti-satellite (ASAT) attacks, however, continue to increase in number and the size of this debris also has also mounted in recent years. Although some graphs show a leveling of debris due to increased due diligence against explosive bolts and controls on fuel tanks and de-orbiting, other projects suggest that break-up of larger debris could start an "avalanche effect" that could greatly increase orbital debris. At this point we have competing models as to what the future holds.

One source of debris is discarded hardware. For example, many launch vehicle upper stages have been left on orbit after they are spent. Many satellites are also abandoned at the end of their useful life. Another source of debris is spacecraft and mission operations, such as deployments and separations. These have typically involved the release of items such as separation bolts, lens caps, momentum flywheels, nuclear reactor cores, clamp bands, auxiliary motors, launch vehicle fairings, and adapter shrouds. Material degradation due to atomic oxygen, solar heating, and solar radiation has resulted in the production of particulates such as paint flakes and bits of multi-layer insulation. Solid rocket motors used to boost satellite orbits have produced various debris items, including motor casings, aluminum oxide exhaust particles, nozzle slag, motor-liner residuals, solid-fuel fragments, and exhaust cone bits resulting from erosion during the burn.

A major contributor to the orbital debris background has been object breakup. More than 124 breakups have been verified, and more are believed to have occurred. Breakups are generally caused by explosions and collisions with other objects in space, but the majority of breakups have been caused by explosions. These can occur when propellant and oxidizer inadvertently mix, residual propellant becomes over-pressurized due to heating, or batteries become over-pressurized. Some satellites have been deliberately detonated. Explosions can also be indirectly triggered by collisions with orbital debris.

Three collisions are known to have occurred since the beginning of the space age. In addition, the debris research community has concluded that at least one additional breakup was caused by collision. The cause of approximately 22 percent of observed breakups is unknown. Approximately 70,000 objects estimated to be 2 cm in size have been observed in the 850-1,000 km altitude band. NASA has hypothesized that these objects are frozen bits of nuclear reactor coolant that are leaking from a number of Russian RORSATs, although this has not been conclusively determined.

At altitudes of 2,000 km and lower, it is generally accepted that the man made orbital debris population now dominates the natural meteoroid population for object sizes 1 mm and larger. The issue of greatest concern is that the man-made debris continues to increase and that an avalanche effect, once triggered, could make the problem much worse.[6]

A series of computer models with names like EVOLVE, CHAIN, CHAINEE, Nazarenko, etc., have been developed by NASA, the European Space Agency, a German university, the Russian Federation Space

Agency (Roskosmos) and others. All these programs seek to project future trends in debris buildup and the likely number of collisions that might be expected to occur in coming years. These models have used different assumptions and different experimental results, but despite variations the conclusions are generally parallel. The major finding is that the growth of debris is expected to accelerate unless there is a halt to space launches and improved controls are placed on satellite operators to limit collisions that are the major source of debris. The problem is clearly most severe in low earth orbit below 2000 kilometers.[7]

The threat is thus greatest to satellite and space exploration operations in low earth orbit, rather than for space tourism flights for sub-orbital flights. The projections, however, suggest that by the middle of the twenty-first century the problem will only get worse as more mass is launched into orbit, especially if ASAT missiles were to explode more satellite systems in low earth orbit so as to create an exponential increase in debris. Explosions of batteries and fuel tanks are a serious concern as well.

Figure 8.1: NASA's Long Duration
Exposure Facility (LDEF)
(Photo courtesy of NASA)

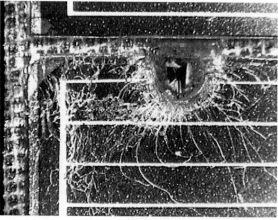

Figure 8.2: Astronaut's Photograph of Damage to a
Solar Cell Panel of the Hubble Telescope
(Photo Courtesy of NASA)

The hazards from debris, from meteoroids, Leonid micro-meteorite showers, solar wind, and other natural sources such as radiation, are a hazard to space travel that at least is measurable and relatively predictable. The dangers that arise from man-made debris, however, continue to expand. NASA has attempted to explore these phenomena by deploying a satellite named the Long Duration Exposure Facility (LDEF). This test satellite was recovered after a five and half year mission to measure impact damage. This satellite (See Figure 8.2) documents that the amount of debris continues to expand at an alarming rate. When retrieved, the LDEF was covered by more than 30,000 craters visible to the naked eye. Of these impact craters, 5,000 had a diameter larger than 0.5- mm. Debris of 1-mm is sufficiently large to endanger an astronaut on a space walk, and can even damage the wind shield of a Space Shuttle. The European Retrievable Carrier (EURECA), the Japanese Space Flyer Unit, and elements of the Hubble Telescope solar array have all been recovered from space. These all show similar levels of damage. The same is true with regard to the Space Shuttle windows and its external surfaces. The most pessimistic projection suggests that if current space launch operations continue, by 2025 there will be 10 major collisions that will increase to over 50 major collisions by the end of the century – a result that will create a deadly hazard to low earth orbit operations.

The chances of an orbital debris-occasioned accident occurring during a space plane's short-term parabolic ascent and descent on a sub-orbital flight are today considered remote. The science of projecting the probability of an accident is at best problematic, given the fact that the various space agency and university computer models resulted in rather significant differences as to the projected number of collisions. Because most sub-orbital flights are not expected to fly higher that 120- kilometers and the greatest concentration of orbital debris is in the 500- to 2000- kilometer range, the possibility of obtaining insurance coverage against such a remote hazard seems viable and at a cost that would be affordable, even for $20-million of protection against the loss of a space plane and its passengers. This insurance would likely also provide coverage against orbital debris, and also protection against any radiation harm to passengers, etc. Again the short-term nature of the flight – with only a few minutes above the protective ozone layer – would render the danger of radiation illness to passengers to be slight. For instance, the radiation levels for astronauts aboard the ISS are considered to be fifty times that of people at the Earth's surface. But the space tourist flight into a similar radiation level would be on the order of ten minutes. The radiation exposure for flight crews on airlines flying regularly over the Polar regions where there is an ozone hole, would, for instance, be higher than that of a space tourist.

What Next?

The space tourism industry excites our imagination and we believe that in time commercial space flight will represent a key step in the evolution of human kind. One may hope that commercial space ventures can move the human race forward to new vistas where the sky is no longer the limit. Space Adventures has already announced plans to send "Citizen Astronauts" around the Moon and commercial stays in space habitats and space walks may be less than a decade away.

The best way for space tourism to succeed, however, is for the above potential "show stoppers" to be addressed and resolved now rather than later when retrofits and retooling of systems can be much more expensive and difficult.

We have gone from thinking of "Citizen Astronauts" as science fiction to tomorrow's fact. But much more needs to be done. We have suggested that attention needs to be given to key issues such as minimizing orbital debris. This will become an increasing problem if we do not solve it now. Even more important is the need to reduce adverse environmental impacts, particularly in the context of the ozone layer and the release of greenhouse gases. At this stage jet airliners are a much greater worry than space planes, but if the number of spaceplane operations increases greatly and hypersonic craft should seek to provide regular transportation services, the environmental problems will only escalate. We need to develop a space insurance capability to allow space tourism businesses to thrive. Most importantly, we need to forthrightly address the dangers of the weaponization of space. A misstep here could not only destroy the space tourism business but also lead to even greater dangers. If deployment of weapons in space should proceed, this could truly lead to World War III.

Finally there is another type of concern and that is whether the US Congress has finished its work on space commercialization? The Commercial Space Launch Amendments Act of 2004, that was adopted in December 2004 and signed into law in early 2005, has the warm and cuddly name of "the CSLAA" among insiders. (In case this is not clear this is a joke.) The CSLAA has been a good basis for the FAA to engage in a thorough and productive rule making process that covers crew and passengers on space tourism flights as well as setting public safety guidelines for ground operations, but the process should not and cannot end there.

The FAA intends to review its rule making based on experience. We have elsewhere recommended a White House Commission that will undertake work that closely parallels its effort with regard to civilian aviation and its safety enhancement. Further a safety hotline, similar to NASA's Safety Reporting System, is also recommended. But this is only for starters.

The issue that should be addressed post 2012 is that of how to eventually separate FAA's duties and responsibilities as "Safety referee" from that of market "cheerleader." It is never a good idea to combine these roles, at least, over the longer term. This is why Congress eventually removed FAA's promotional role from aviation and why at some point NASA should either be in charge of space transportation operations or space safety, but not both.

The haunting question remains: What might have been the outcome for the ultimate Challenger and Columbia Space Shuttle missions if we had just separated NASA's role as space system developer from safety standard enforcer. In this case, the role of Congress in setting the mission goals and then also controlling the purse strings and ultimate schedule could be questioned in terms of their culpability as well.

Anyway, NASA has largely deserted its role as developer of advanced aviation and aerospace technology long ago. The military and aerospace companies have essentially inherited this role although NASA Glenn is still doing some work of significance to develop new jet engines that burn much cleaner. This work, however, does not extend as far as developing hydrogen fueled engines or more advanced R&D. We thus see Congress as "missing in action" when it comes to key questions about setting a clear course for US research and development as well as safety oversight in a number of aspects of space, aerospace and aviation. Lord knows we probably do not need another study commission, but perhaps Congress might task a qualified non-profit study group, that reports directly to Capitol Hill staff, to address such issues as the following:

- Should development of advanced aviation and aerospace technology and safety innovations be transferred from NASA and thus make it only the National Space Administration? (These tasks along with the Langley Research Center might be, for instance, transferred to the Department of Transportation along with a viable budget.)
- Should the FAA continue to be charged with regulating the safety of space tourism while at the same time being asked to promote its development? If not, should these functions be moved to the Department of Commerce or at least be moved to two different offices within the FAA or separated even further apart?
- Is NASA just responsible for developing new rocket technology? There are exciting new possibilities that might provide lower cost, safer and more environmentally sound technology to lift humans and payloads into space. These include tethers, space elevators based on nano-tube systems, solar sails, nuclear propulsion, advanced ion engines, and even lighter-than-air craft. It seems that in many ways NASA strategic plans are flawed and unbalanced in terms of space versus advanced aviation. NASA's goal, at times at least, seem skewed to the interests of powerful political and corporate forces, as well as an unimaginative approach to developing truly breakthrough technologies. NASA's only attempt to think outside of the box and envision truly new technology was known as the NASA Institute for Advanced Concepts (NIAC). But this was shut down due to the all too familiar budgetary constraints. Such efforts probably need to reside within the National Science Foundation (NSF) or a Civilian Advanced Research Projects Agency if it were to be resurrected in future years.
- Further NASA's sprawling and decentralized research centers sometimes end up overlapping in their missions and remain unavailable to support other parts of the US Government. Reforms could start with making these centers into Federally Financed Research and Development Centers (FFRDCs), as recommended by several previous study commissions, including most recently the so-called Aldridge Commission.
- To state the obvious, the ITAR regulations need to be revised and streamlined. The process today is cumbersome, expensive, and ineffectual.

The system clearly has shot US aerospace industries in the foot by making it impossible to compete for many international contracts for communications satellites, remote sensing satellites and other aerospace research awards. It was said by McGeorge Bundy that if you have a process that protects diamonds and toothbrushes with equal vigor, then you will probably loose less toothbrushes and more diamonds. Certainly ITAR needs to be revised to protect the diamonds and worry a lot less about the strategic value of toothbrushes.

This is just the start of what might turn into a laundry list of issues. We hope the focus can remain on a new US vision in space that can look a couple of decades ahead.

Congress should recognize that all is not well or right in the world of NASA, the Department of Transportation or the FAA. US legislators should attempt to work out some of the key "wrinkles" that have emerged in the fabric aerospace research and development. US Agencies that address space and advanced aviation and aerospace technologies should be given new direction, new goals and in some cases hands need to be untied. Corporate innovation related to space planes, new space and aviation transport systems, etc. should be strongly encouraged, while also addressing serious environmental, public safety and regulatory issues.

NASA and the Department of Transportation (DOT) need to be gone over with a "policy iron" that smoothes its "wrinkles" and restores leadership to US space and aviation programs. In short, Congress and a new President need to work together to create a national space agenda that is a whole cloth and not a raggedy patchwork kluge sewn together by special interests and short term thinking.

References:
1. TED Case Studies, SST Pollution – http://www.american.edu/ted/SST.HTM
2. Joseph N. Pelton, Space Planes and Space Tourism, (2007) – George Washington University, Washington, DC. P. 5 – www.spacesafety.org
3. Joseph N. Pelton, Space Planes and Space Tourism, (2007) George Washington University, Washington, DC. Appendix 3, www.spacesafety.org
4, National Security Policy Directive of August 31, 2006 on US Space Policy – http://www.fas.org/irp/offdocs/nspd/space.html
5. National Security Policy Directive of August 31, 2006 on US Space Policy, http://www.fas.org/irp/offdocs/nspd/space.html
6. "Orbital Debris," Center for Orbital and Re-entry Debris Studies – www.aero.org/capabilities/cords/debris-basics.html
7. Technical Report on Space Debris – United Nations Committee on the Peaceful Uses of Outer Space, 1999 – http://orbitaldebris.jsc.nasa.gov/library/UN_Report_on_Space_Debris99.pdf

"The military potential of manned spacecraft may remain an unresolved question for a long time."— Maxime Faget

The Strategic Elements of Space Plane Systems

Why Space Planes?

There are many reasons why space planes might be developed. These include space tourism, hypersonic transportation, space science experiments, testing of new materials and systems and low cost launch of small satellites. Then there is national security. The private sector can and probably will develop space planes to serve all of these potential "markets," but governmental funding related to space plane programs, not too surprisingly, will focus largely on national security and defense-related applications.

However, as we have seen in the previous chapters, most private ventures are currently engaged in developing systems for space tourism.

NASA's various ill-fated space plane projects have now all been canceled. The US civilian space agency abandoned the development of space planes as a prime objective in 2004 when it refocused on missions to the Moon and Mars. The X-37 space plane (or Orbital Test Vehicle) program has been transferred to the US Air Force (USAF) and the Department of Defense (DOD), although NASA continues to play a minor role in this program. In addition, the X-43, after achieving a record speed of Mach 10, has also been discontinued by NASA and moved to DOD. In short, NASA has redirected its efforts away from space plane projects. To this, many would say 'hooray'. After years – actually make that decades– of unsuccessful efforts to develop one version or another of space plane technology, NASA finally gave up. Failures, or at least discontinued efforts, included the HL-20, X-33, X-34, X-35, X-37, X-38 and X-43. Ironically, NASA had defined developing space plane systems as an essential part of its mission up until 2003, but then turned on a dime to say: "Oh no, this is a mission for someone else." The US military, long the strong proponent of a space plane to orbit and back as a prime objective, quickly moved in to take over.

The Strategic Use of Space Planes

During the George W. Bush administration the strategic aspects of space systems and technology mushroomed in importance. On one hand, NASA was asked in 2004 to pursue the vision of exploring outer space and mounting a mission to the Moon and Mars. On the other, the DOD was asked by the White House to take over developing space plane systems and near-Earth military space applications. The nature, purpose and scope of US defense space plane development programs, however, are not entirely clear. In fact, because some of the programs are classified, no one is entirely sure what capabilities have been developed. For instance we don't know which space plane systems are to be robotically controlled and which are capable of being manned and perhaps also remotely controlled.

There are ongoing efforts around the world to develop space plane technology that may lead the way towards space tourism goals in the private sector, but parallel governmental programs seem to have strategic objectives as well. In short, all space- faring nations, at least the US, Russia, China, Japan and France, probably have as one of their objectives the improvement of their strategic space capabilities and related space plane technology. The potential strategic objectives for space plane projects certainly could include many different applications. These might, for instance, include better space reconnaissance systems, better communications and networking capabilities as well as their rapid deployment, and even anti-satellite (ASAT) systems or even the ability to deploy small micro-satellites.

There is, of course, even the possibility of future deployment of military weapons systems in space or the use of space planes to disable space-based hostile laser systems, or other defensive actions against offensive space weapons. A space plane might also be able to deliver weapons of various types and forms. Unlike a ballistic missile, a space plane could change its orbital trajectory and be able to disguise its entry path for the delivery of a weapon system such as a bomb. It could also be configured to deliver weapons at extreme velocity that could, on impact, destroy a targeted underground and fortified bunker by deploying so-called "rods from god."

The Deadly Start of Space Weaponization – in 1944

There is an interesting and deadly precedent here dating back to World War Two and the German rocket program developed by Werner van Braun. Some 3,000 V2 rocket bombs were launched against targets in the U.K. and Belgium. These weapons were Hitler's final attempt to reverse the outcome of the war by halting the advance through Europe and terrorising the population of Great Britain.

These rocket bombs, fired from launch sites in Germany, Belgium and Holland, each carried a warhead of 2150 lbs of Amatol, reached an altitude of 55-miles (88.9-Km) and had a range of some 200 miles. Over 6,000 of them were built by Germany's forced labor workforce, and over 3,000 were actually launched – 1,402 of them targeting the U.K. Accuracy was a considerable problem but the records show that these bombs from outer space caused the deaths of 2,700 civilians and more than 10,000 injuries. Winston Churchill's War Memoirs make it clear that he was seriously concerned about this threat after learning the details from an intelligence source in 1943. This was when von Braun and his team gave a secret demonstration to Hitler at the Peenemunde research establishment in Northern Germany. Hitler was so impressed that he gave the order for 10,000 V2 rockets to be built ... a way, perhaps, for him to change the course of the war.

In London, Churchill's war cabinet recognised that this was a new weapon for which there was no defense, but with the potential to seriously damage morale. The only defense was to attack the factories producing the V2's and supplying the rocket fuel – and to destroy the launch sites then being constructed. The RAF began a series of risky low level bombing raids targeting the main factory at Peenemunde, at a heavy cost in aircraft and their crews. Following the D-Day landings in June 1944, the V2 threat influenced the war strategy so that the direction of advance went northwards from Normandy to neutralize the operational launch pads. Hitler responded by moving the production and launch facilities back into Germany and regular V2 attacks continued on South East England and on the Belgian port city of Antwerp after it had fallen to the Allies; and some rockets also hit Paris after it had been liberated.[1]

The U.K. had already experienced the pilotless V1 bombs – known colloquially as 'doodlebugs' or 'buzz-bombs' – and Londoners became accustomed to listening for their unmistakable engine noise, and then taking cover when the aircraft went silent. It was not until September 1944 that the first V2's began to fall on London, but Churchill waited until November before making a low-key statement to Parliament about 'the German rocket attacks'. By then, the Allied advances in France and the Low Countries towards the Rhine and Ruhr were making the headline news but rockets continued to fall on Britain until March 1945 when the last recorded V2 bomb hit Croydon, just four months before VE day.

As the war moved towards its close, the progress made by the German rocket industry was of considerable interest to the Russians and Americans who both sought to retrieve equipment and personnel to assist in their own space research programs. In fact, Werner von Braun and his team surrendered to the US forces rather than fall into Russian hands. And von Braun himself subsequently played a major role in the development of America's manned space program.

In the Cold War environment that ensued in the post World War II period, the German rocketeers "rescued" by Soviet forces became a key component in the Soviet team that developed rockets and atomic weapon missile systems for the U.S.S.R. This was mirrored by the US efforts led by Werner von Braun to develop ICBMs to deliver American atomic weapons. Space systems of all types tended to become weapons systems. Missiles were developed to deliver explosives and even nuclear devices. Communications systems were used to aid strategic communications. Spy satellites were developed for espionage. Geodetic and space navigation systems were used to target bombs and for strategic mapping and instant location determination for blue and red forces.

As the Cold War heated up, or should we say became more frigid, there was an ever-increasing proliferation of space weaponry to support the incredible strategy of MAD (Mutual Assured Destruction). For decades, ever more sophisticated missile systems served to keep American and Russian populace in a constant state of apprehension. New technologies evolved such as Multiple Intercontinental Re-entry Vehicles (MIRVs), anti-ballistic missiles (ABMs). Then came the attempts to create a "shield" over an entire country against nuclear missiles, known by warriors as the Strategic Defense Initiatives (SDI) but known mostly to the public as Star Wars.

With the fall of the Berlin Wall and the demise of the U.S.S.R., space-based MAD-driven nuclear tipped missiles began to be decommissioned and from the 1990s the world began to breath a bit easier. Despite these changes, strategic uses of space continue to be a part of the modern condition. Although there have been calls to create a treaty that would "outlaw" space weaponisation, military experts claim that such a ban on weapons is not verifiable. Further there is a gray line between when a communications satellite, a remote sensing satellite, a laser system or a steerable space vehicle "is or is not" a space weapon. Some have suggested that a Space Shuttle could be used as a space weapon. Certainly a space plane designed for sub-orbital flights for space tourism could also be used for different applications, as both an offensive or defensive space weapon. For these reasons governments are likely to regulate the use of space planes and US ITAR trade restrictions will likely apply to the sale of space planes or their components.

Onwards to the 21st Century and "Project Constellation"

Space based systems, whether commercial or defense-based, are becoming more and more sophisticated and diverse. There is clearly a growing number of potential defensive or offensive applications for space systems that go beyond the delivery of explosives, nuclear devices or bio-chemical weapons. There are indeed defense-based programs that are a combination of space and other systems such as the so-called "C-2 Constellation" project of the USAF. This project has as its goal the integration of space assets, linking them together instantly or at least in "near real time" with sensor, navigation and communications systems. Such a project would interconnect space, air and ground sensors with so-called "Netcentric" communications networks that extend right out to the edge of military operations where actual hostilities may be occurring. Thus Project Constellation would hook up and integrate incoming and outgoing data via high-speed computers. Countless sensors of all types and at all locations around the world would be available for military operations, regardless of whether the sensor is located on land, at sea, on board aircraft, or even aboard Unmanned Automated Vehicles (UAVs) or satellites.

The various elements of Project Constellation would, over time, grow to include the Mobile User Operational Satellite (MUOS) system and the super-fast laser-based satellite communications known as the Transformational Satellite System (TSAT) that would encircle the globe. All of these new satellites are currently planned for orbital insertion using conventional rockets, but space planes might allow future elements of such a system, such as space sensors to be deployed and serviced at lower cost.

The deployment of these "netcentric" warfare systems by the US and its allies depends on space sensors, space reconnaissance, and space communications networks.

The many layers of systems for acquiring information, communicating to weapons systems, and processing data are represented in Figure 9.1 above. Clearly space planes, once developed, will be utilized for a number of functions and increasingly become a part of the strategic planning process. Only because the X-43 and X-37 space plane programs were initiated by NASA did the public have visibility into these programs, but these are now classified projects. While they were still unclassified projects, these vehicles have been tested to speeds on the order of Mach 10.

Figure 9.1: Defense & Strategic Systems Are Now Dependent on "Smart Weapons Systems – Will Space Planes Soon Be A Part of the Mix?

The use of space-based communications relays, space sensors, and anti-satellite weapons (i.e. ASATs) are part of today's reality. Perhaps, in time, we may see space planes or very high altitude UAVs with many types of capabilities that include advanced communications, reconnaissance and surveillance. Others could be used to deploy on an instantaneous basis "theatre communications systems" associated with a particular geographic hostility. Most significantly we might potentially see space planes armed with what might be called space weapons. Crewed and highly maneuverable space planes armed with various types of weapons would likely change the strategic dimension of space systems and certainly create problems and complications for commercial space enterprises.

As discussed earlier, the latest US announced space policy, as released from the White House in August 2006, has elements that seem internally inconsistent with other elements of the same pronouncement. It also seems as if different parts of the US Government, such as the DOD, the National Security Council and the Department of State, might have drafted different elements of this official statement of policy. On one hand this statement clearly states that the United States is: "committed to encouraging and facilitating a growing and entrepreneurial US space sector" and indicates it will use the commercial space sector to the maximum extent possible, consistent with national security. But on the other hand, it indicates an increased focus on the strategic importance of outer space. This Directive, as cited in chapter 8, sets forth at least two principles that suggest that the US views the uses of space in both a security and military context, and first and foremost in a strategic context. Such a stance would seem to make it at least questionable for commercial organizations to invest in "space deployed infrastructure" or even orbital tourism enterprises if they feared that their high expense and high value investments would be subject to military attack or appropriation for military or national defense purposes.

On the other hand, it is also clear that of the US commercial enterprises pursuing the development of space planes or access to orbital systems, close to half are either funded or being otherwise encouraged in their efforts by either NASA or the US DOD. The current burning question is whether there will be a new Presidential Directive from a new administration in 2009 that reverses or at least alters the August 2006 statement?

As described in the previous chapter, there are essentially three types of enterprise that are pursuing what might be generally characterized as the "space tourism" or the "commercial access to space" market. These are the major aerospace corporations, which have traditionally been heavily funded by defense as well as NASA civil contracts; the mid-tier aerospace corporations, which often support the major aerospace corporations in many defense-related contracts; and then the true start-up companies that may also find that their future success depends on governmental contracts and support from the defense sector. The bottom line is that perhaps half of the money for the space plane industry comes from defense budgets – not only in the US but abroad as well.

Non Commercial Space Plane Projects from the Military Sector
Currently the US DOD has at least two space plane development programs that are active. These include the X-37 system that was jointly developed by NASA and the USAF, but turned over to the DOD in late 2004 when NASA abandoned this project to focus its efforts on the Moon / Mars project.

The Air Force is currently working on an unmanned space plane that is directly based on NASA's X-37 program. This craft was at one time planned to be the direct follow-on to the Space Shuttle. If successful, this spacecraft would be the first since the Shuttle capable of returning experiments back to Earth for analysis. What is unique about this particular development is that it has been redirected to operate robotically for sustained periods of time, with missions that last up to nine months in orbit. This reusable Orbital Test Vehicle (OTV), as now designed, would be about one-fourth the size of the Space Shuttle and would deliver objects into low-Earth orbit in its experimental bay, much like the Shuttle's payload bay except smaller. The OTV could then continue orbiting for months before bringing the objects back to Earth. This would allow it to test how satellite components react to long stays in space, among other tasks. The OTV can and will also help test technologies for more advanced reusable space vehicles.

The OTV's first flight was scheduled for some time in 2008, launching on an Atlas 5 rocket from Cape Canaveral Air Force Station in Florida. The first one or two flights will not carry any experiments. Instead, they will test the craft's ability to autonomously re-enter the atmosphere and land, as well as probe new landing gear and lightweight structures that can withstand the heat of re-entry. The OTV is designed to land

Figure 9.2: The U.S. Air Force's Orbital Test Vehicle is Being Designed to Make Sustained Mission Round Trips to Space
(Photo Courtesy of the U.S. Air Force)

in California, either at the Vandenberg Air Force Base, which hosts expendable rocket launches, or Edwards Air Force Base, which acts as an alternate landing strip for the current Space Shuttle.[2]

The OTV overall management is provided by the Air Force Rapid Capabilities Office. There is also program collaboration with the Air Force Research Laboratory, the Defense Advanced Research Projects Agency (DARPA) and NASA. NASA's role is, however, largely defined by handing-off its institutional knowledge from the preceding X-37 program.

Boeing, not too surprisingly, serves as prime contractor for the OTV program. This selection of course stems from the fact that Boeing was the commercial lead for the NASA X-37 technology demonstrator. There is accordingly a significant amount of continuity from the X-37 to the OTV program. The OTV is nevertheless a new program and there is new technology involved with this development that represents much more than simply a name change.

According to a statement from the Secretary of the Air Force, the OTV program will focus on "risk reduction, experimentation, and operational concept development for reusable space vehicle technologies, in support of long term developmental space objectives." This is the rather vague statement that is often associated with a classified military program and thus the actual use of the OTV could be for almost any purpose.

In just a few more years, there should be a range of US space vehicles that would have the ability to return experiments to Earth. While the OTV is unmanned, the new Orion vehicle that NASA is developing to go to the ISS and the Moon will be able to go to and from orbit. Likewise, the craft being developed under the NASA COTS program (i.e. the Commercial Orbital Transportation System that Orbital Sciences Corporation and Space X are actively developing under NASA contracts) will be able to take crew to and from the space station shortly after 2010.

When the X-37 was a NASA civilian space program in the late 1990s, it was to be the first of a planned series of flight demonstrators that NASA called Future X. At the time the X-37 was described as a robotically controlled, autonomously-operated vehicle designed to conduct on-orbit operations and collect test data. Its special thermal blanketing system and shape was designed to withstand the super-hot temperatures associated with re-entry from earth orbit by deceleration from speeds as high as Mach 25. This is a velocity of around 26,500-kilometers per hour or about 18,000 miles per hour.

The initial plans for the X-37 envisioned that it might be lifted to orbit by either a Space Shuttle or a launch vehicle like the Atlas or Titan and would then be deployed in Earth orbit. The vehicle would fly through space for some period of time and perform a variety of experiments before reentering the atmosphere and landing on a conventional runway. NASA transferred its X-37 technology demonstration program to DARPA in late 2004 without the completion of this development program. This was but one of the Future "X" programs that was interrupted and canceled by NASA, in part due to delays and cost overruns and in part due to the high cost demands of the Space Shuttle program and the drain of the Space Station.[3]

NASA had earlier planned on two X-37 unmanned vehicles. One of them, which was actually built, was known as the Approach and Landing Test Vehicle (ALTV). The other craft, that was designed but not built, was known as the Orbital Vehicle. The purpose of the ALTV was to test re-entry trajectories from within the atmosphere for possible later use on the more advanced Orbital Vehicle. This vehicle demonstrator, that was "inherited" by the DARPA from NASA in 2004, was actually tested in April 2006 with less than ideal results.

In the test, the vehicle was released from Burt Rutan's White Knight very high altitude carrier aircraft. The White Knight is, of course, best known as the aircraft that carried the SpaceShipOne to high altitude from where it blasted its rocket engine to win the $10-million Ansari X Prize. The ALTV version of the X-37 ALTV, however, was not able to maintain course in landing and ran off the runway.

The first one or two flights of the OTV, probably in September 2008, will not carry any experiments. Instead, these flights will be tests of the vehicle's ability to re-enter the atmosphere under remote control and land successfully. Other test objectives will be to prove the functioning of new landing gear and of the very lightweight thermal materials designed to withstand the super hot heat of re-entry.[4]

The X-37 is not the only governmental program to develop space plane technology that NASA was passed on to the military. The other program, known as the X-43, is designed to prove the viability of so-called scram jet technology.

X-43 – Another Future X Program Discontinued by NASA

This is the first truly successful "air breathing" rocket engine to provide sufficient oxidizer via intake air flow to fuel a continuously firing rocket engine at multi-Mach speeds. This program is now proceeding on a classified basis. The spacecraft takes off using a B-52 aircraft, which then drops at altitude a modified Pegasus launch vehicle that accelerates the X-43 to sufficient speed that its scram jet motor can operate and achieve flight and acceleration to speeds of very near Mach 10.

This hypersonic craft was initiated by NASA in partnership with DARPA. NASA undertook to develop at least three X-43A craft for testing. DARPA also agreed to design in parallel its own version of a scram jet hypersonic vehicle known as the X-43B and X-43C to various scales. After successful tests by NASA the X-43A reached nominal test speeds of Mach 10 in 2004. NASA turned over the subsequent development to DARPA after these tests were completed.

In addition to the X-43, the DOD is also carrying out tests of other spaceplane systems that include prototypes developed by Lockheed Martin, Boeing, Northrop Grumman, Orbital Sciences, and others. Further, there were dislosures in 2006 in *Aviation Week and Space Technology* (sometimes aptly nicknamed, the "Aviation Leak") that there might be a now discontinued two-stage-to-orbit space plane operated by the USAF. This unverified vehicle was supposedly designed by Boeing so that the first craft would fly to very high altitude and the second stage could then be "rocketed" into low earth orbit. This elusive and perhaps entirely mythical craft called the Blackstar (also known as the SR-3/XOV) is, in theory, a now moot

Figure 9.3 The X-43 Unmanned Scram Jet Technology Demonstrator
(Photo Courtesy of NASA)

proposition, since the Aviation Week story claims that this secret program (if it ever existed) has been shut down, either for cost or performance reasons.[5]

Certainly the existence of such a plane has been considered by experts as "unlikely" for a variety of cost and technical performance reasons. The X-47A, however, apparently is quite real, as are other classified high altitude systems (see illustration at: http://www.physicsroom.org.nz/log/archive/15/spaceplanes/). These various defense-related developments have also led to potential spin-offs such as Lockheed Martin's new QSS (i.e. Quiet Super Sonic) prototype for a hypersonic personal executive jet for transoceanic service.[6]

For a number of years, there has been tension between NASA, as the civilian space agency, the USAF, the US Congress and the US Executive Branch (especially the Office of Management and Budget) as to which agency should develop space plane technology. The solution to this problem seemed to have emerged with the so-called Moon / Mars Vision Statement in 2004. This Presidential directive assigned the top priority to the US space agency to develop space systems to go to the Moon and Mars. The bottom line of this Moon / Mars Initiative is that NASA has turned over the space plane initiatives (to wit the X-37 and the X-43) to the USAF and DARPA.

The other consequence of this change in direction for US space policy has been less overt. This is the further guidance from the White House for NASA to phase out the Space Shuttle and its participation in the International Space Station (ISS) as soon as possible. The completion of the ISS is at least a decade behind schedule at this point, and the experimental mission for the Space Station is seriously in question. Such directives appear to make a great deal of sense, in light of these rather dreary facts concerning budget and schedule. Ironically, Japan, Europe and Russia seem well positioned to make use of the ISS for experiments and as a way station for going into space.

There is now, at least to some, an almost "unwelcome clarity" in US space policy that did not exist prior to 2004. This redefined policy largely ends up with NASA being much more narrowly focused on Deep Space and the business of exploring the "great out there" and thus not pursuing "missions to planet earth" and a reduction in space science initiatives. Environmental monitoring or developing applications satellites are not high on the Bush White House agenda.

Clarity, however, is not necessarily a good thing. A number of scientists – particularly those concerned about the US role in commercial space applications, global warming, and atmospheric and meteorological concerns – feel that this redefinition of priorities leaves important civilian space programs adrift and under-funded. This is certainly true if one compares how NASA spends its resources with other national space agencies. While the agencies of other countries are much more focused on space applications and the use of space systems to help people here on Planet Earth, NASA and the US administration, particularly during the George W. Bush presidency, has said they will leave that to other space agencies, universities and private enterprise.

For better or worse, this is how space priorities have been redefined in recent years. However, the US Presidential elections, increasing tensions in the US and Russia over goals and objectives—on earth and in space—as well as recent schedule and program delays with the Ares I and V launchers could certainly end up redefining the Constellation Moon Program. The next few years and a new President in the US may well redefine goals and timetables for the ISS and for, strategic objectives in outer space as well as commercial space initiatives. Priorities could change in a new US administration, but at present, the impetus toward implementing the Orion and Ares vehicles to go to the Moon represent Job One for NASA. This means leaving space programs in low earth orbit and for "near space" more and more in the commercial sector.

Many see NASA's abandoning of space plane development as both a logical and desirable decision. They ask: "Why should NASA fund development of space plane technology when the DOD has the funds and is willing to take over this development?" The Space Exploration Foundation has said, rather cynically, that the Air Force and DARPA can certainly do no worse than NASA which has unsuccessfully undertaken a half dozen space plane projects and terminated all of them when they were all behind schedule and well over budget. NASA applications for a space plane are limited, especially since it has now awarded its COTS research contracts to Space X and Orbital Sciences. Both of these contractors, however, are also receiving funds from the US DOD. Space X is funded by the Air Force to develop the Falcon launcher to launch defense satellites at lower cost, while OSC has received funds from DOD to build both satellites and launch systems. Likewise XCOR, in announcing its new Lynx space plane in 2008, released information that it has a contract with the USAF to share information with regard to the testing and performance of this new vehicle.

Human Control Versus Robotics
The potential applications of space planes for strategic purposes have been variously listed as: (i) fast response reconnaissance vehicles to emergency conditions (including natural and man-made disasters as well as dynamic and critical war conditions); (ii) rapid response and elusive systems for delivery of military strikes and bunker penetrating projectiles (i.e. "Rods from God"); (iii) quick deployment of small satellites on an on-demand basis; (iv) other applications that include research, testing of materials, and re-supply of a variety of different application satellites.

There are several key debates going on at present within the strategic defense community. One is over whether it is possible to use space for military purposes under current space treaties. Some argue that one can use space for offensive military strikes if you believe you are under threat of attack. Others argue that one can only employ defensive space military weapons such as anti-missile systems to create a protective shield. Yet others argue that no military use of space is appropriate or even "allowable" under space treaties.

Another new debate is emerging within military circles, not over the existence of space weaponry, but rather how it is controlled. This debate is between those who believe that military space planes and like technology should only be designed to operate under human control and those who argue for robotically controlled vehicles. The advocates of robotic systems maintain that such space weapons systems are more efficient, less costly and even more accurate in achieving their various potential objectives. Many would argue that having pilots in the cockpits of space planes is an overrated concept. It might look exciting or even heroic in a "Star Wars" movie, but from the standpoint of cost, performance and effectiveness, robotic operations could potentially make more sense (and much less dollars and cents).

An article by Major David M. Tobin of the USAF presents in great detail the reasons why the cost of providing life support systems for humans in space is enormous and that computers, avionics and robotic control systems can often perform strategic missions better and at lower cost. It is a little known fact that the Space Shuttle could have been designed to operate without a crew and that a proposal to replace the Challenger with a robotically controlled Shuttle without a life-support system was seriously considered. This would have made the Endeavor a much lower cost vehicle and had other Shuttles been retrofitted for robotic operations the tragic loss of life with the Columbia might never have happened. NASA, however, over the last fifty years has consistently promoted "manned" space programs because top US space officials know from detailed public opinion polling that astronauts hold human interest while "expendable" machines, in the form of robots, do not. Scientists from the Planetary Society have argued for years that we could find out much more about the solar system for much less by using remotely controlled machines that do not have to breath oxygen or require expensive life support systems.

In the case of military space planes, there are clearly arguments that can be made on both sides of the "crew vs. automated controls" debate for such vehicles. It is significant that at this point both the OTV and the X-43 scramjet projects are currently unmanned programs. In the carefully analyzed and argued paper by Major Tobin, the conclusion is as follows:

> "A first-generation Military Space Plane (MSP) could function without a man on board—but whether it operates autonomously or under the close supervision of ground controllers remains to be seen. This first- generation MSP could execute at least a portion of all four space-mission areas. It could over fly any point on the planet to deliver a strike payload or conduct a reconnaissance mission. On a counter-space mission, it could destroy hostile satellites using kinetic-energy projectiles or directed-energy beams. As a reusable launch vehicle, it could perform a simple yet critical space support mission—satellite deployment."[7]

Many factors support the development of a first-generation MSP without men on board. First, it could satisfy the near-term mission requirements surveillance / reconnaissance and defensive counterspace – as well as perform at least a limited role in all four space-mission areas. As the less expensive alternative, it stands a greater chance of being funded. Finally, the absence of a crew, their life-support equipment, and a dedicated cockpit help reduce the vehicle's operating weight. Given the technical challenges involved with single-stage-to-orbit flight, any opportunity to reduce the vehicle's mass is advantageous.[8]

There are two pathways forward with regard to military space planes. One involves space policy and international law and the other involves technology. There is no doubt that new technology will bring forward new capabilities with regard to military space planes – both of the manned and unmanned kind. The key, however, is whether new international agreement will limit the use of space for military purposes, whether in an offensive or defensive mode. The potential is certainly there for nations to agree to "de-militarize space," but in the new era since 9/11 the further question that remains is 'what about terrorist groups?' Who knows when techno-terrorists that might attain space capabilities?

The strategic future of space, however, involves more than just military use of space planes. Today there is also rising concern about space debris and its potential to have a very destructive impact on all types of private space commerce. The alarming rise in space debris in time could be a threat to telecommunications, remote sensing, navigation, and meteorological satellites as well as military satellites for surveillance and weapons monitoring (as outlined in chapter 8).

We also need new ways to address the strategic issues raised by the increasingly sophisticated space program. This may mean the creation of a new entity to address such issues as the international regulation of space tourism and related travel through the air space of various countries. Likewise we need more effective international control of orbital debris. Certainly we need a global capability to monitor near earth objects as well as comets and asteroids that might pose a hazard to humanity. Further, it would be a good idea to figure out a way to implement the de-militarisation of space and effective regulatory control of space systems of all types. So far the United Nations' ability to address effectively such space issues has been limited at best.[9]

Albert Einstein once said that the hard part of any issue is finding the right question. At this point the right question seems to be how can we organize the world's affairs so that space-related issues and concerns can be better managed and controlled – especially to avoid star wars and make it safe for civilian passengers.

Other Space Programs – for Warfare or Prosperity?

It is likely that military space plane programs are currently active in other countries around the world. These could well be underway in China, Russia and Europe plus possibly classified programs in other countries. At this time there is no specific unclassified information available on such programs. The European or Russian aerospace firms that would be involved in developing such a capability would, in any event, be the same companies that are involved in developing new space plane capabilities for civil governments or private space tourism. There is also widespread interest in the concept of developing reusable space planes that could operate robotically. From a safety view point, the design of vehicles that could be operated via ground control commands would appear to be an attractive safety feature in the event of the pilot (or copilot) of a space plane becoming incapacitated.

The escalation of warfare into outer space – whether manned or unmanned –should be a major concern to all. As new technology evolves it will be more and more difficult to put the genie of sophisticated space weapons back into the magic lamp of world peace. Today, the Aladdins that can rub the magical lamps of scientific research and produce increasingly menacing space weapons are considered allies. But in another decade or two, these weapons can find their way into the hands of terrorists and rogue nations and hold global society hostage.

There have been extremely foolish thoughts in the past suggesting that we can restrict the "nuclear club" to a handful of responsible nations. This has proven to be wishful thinking. Seeking to develop space vehicles that can deploy a plethora of weapons of mass destruction is equally misguided. The best way forward for our troubled world is not more and more weapon systems – particularly space-based weapons. No, the best path to stability and reasonable tranquillity will involve better global education systems, economic development, and ecologically responsible practices. In all of these areas, space systems offer economic efficiency, effectiveness and equity. A space race to create better and better weapons in the stratosphere and beyond will ultimately end in despair.

It is ironic that as "billions" are being spent to pursue destructive wars around the world, it is difficult to get appropriations for "millions" to support such projects as the Global Legal Information Network (GLIN). This is an electronic system that some forty nations plus the UN, the EU, the OAS and the Arab League are developing to sustain a worldwide rule of law. It is likewise true that a few million dollars could also help start a global television university modeled after the Chinese TV University that was started in 1986 under Intelsat's Project SHARE (Satellites for Health and Rural Education). This began in Beijing with the support of the Ministry of Education and Central China Television as part of the 20th anniversary celebration of Intelsat. This system is now bringing education, training and health to millions of citizens in remote parts of rural China. India has now deployed its Edusat satellite system that, like China, is supporting over a million students. A similar satellite tele-education system deployed on a global basis could ultimately show that, over the longer run, space tele-education can do more for world stability than space weapons. Commercial space may even make such systems available sooner and at lower cost, if the right objectives were set into motion now. Space can be used to fight star wars or pursue peace and prosperity. We can only hope that we ultimately make the right choice. In a complex world we may find we have to do both.

References:
1. Winston Churchill – "The Second World War," Volume 6 – Cassell & Co, London, 1956
2. Leonard David, "US Air Force Pushes for Orbital Test Vehicle" – Space.com, Nov. 17, 2006 – http://www.space.com/news/061117_x27b_otv.html
3. Joseph Pelton et al, "Space Safety Report: Vulnerabilities and Risk Reduction in US Human Space Flight Programs," George Washington University, Washington, D.C. 2005.
4. Kelly Young, "US Air Force to build unmanned space planes," New Scientist, Nov. 20, 2006 http://space.newscientist.com/article/dn10617-us-air-force-to-build-unmanned-space-planes.html
5. William B. Scott, "Spaceplane Shelved?" Aviation Week and Space Technology, March 5, 2006 - http://www.aviationweek.com/aw/generic/story_generic.jsp?channel=awst&id=news/030606p1.xml
6. "All Sonic, No Boom" – Popular Mechanics, March 2007, Vol. 270, No. 3, P. 64-67
7. Major James Tobin – "Man's Place in Space-Plane Flight Operations Cockpit, Cargo Bay, or Control Room?
8. James Tobin, US Air Force, Air Command Air University, Report Number AU/ACSC/285, 1984-04, Maxwell Air Force Base, Alabama
9. Joseph N. Pelton, Tommaso Sgobba, Nicholas Bahr, and Ram Jakhu – Orbital Debris and Space Security – Space News editorial, June 2007

"I thought it would be hard, and it's harder than I thought. But I want to make rockets 100 times, if not 1,000 times, better. The ultimate objective is to make humanity a multiplanet species. Thirty years from now, there'll be a base on the moon and on Mars, and people will be going back and forth on SpaceX rockets." — Elon Musk, interviewed by Carol Hoffman of Wired Science.

COMMERCIAL SPACE FLIGHT AND THE SWEEP OF HUMAN EVOLUTION

The "Ascent of Man" 21st Century Style

The future of space planes, space tourism, and the destiny of advanced civilization are inextricably linked. It is our contention that space tourism and private space ventures are about far more than cosmic joy riding. Remarkably, the dawn of the space age came a half century ago. To date only about 500 astronauts and cosmonauts have rocketed into space. For better or worse these early space travelers, and the scientists who designed their spacecraft, have pioneered a fundamental shift in the course of evolution for *Homo Sapiens*. As the entrepreneur Elon Musk, the astrophysicist Stephen Hawking, and the Apollo Astronaut Rusty Schweickart have all noted from rather diverse perspectives, the future of humanity requires going into space permanently. If we are to survive as a successful species with a shelf life longer than the dinosaurs we must establish a multi-planet civilization.

It was Dr. Joseph Bronowski who coined the title "The Ascent of Man" in his famed TV series and book in 1973.[1] If he were writing today perhaps he might have thought to entitle his book "The Ascent of Humans" – but that is another story. Bronowski looked at the major steps in mankind's evolution: the transition from hunting and gathering to farming and eventually to building cities. Now the final frontier is to expand human civilization into space. Without the evolution of culture itself, he said, man's potential would never have flowered. He went on to examine why humans have the longest childhood of any species. Indeed scientists are still trying to understand the tremendous potential of the human brain. Why does it take so long to develop? Apparently just as human civilization needs so much time to evolve so does the human brain. Why is it that different parts of the human brain seem to work at odds with other parts? (In the movie *I Robot,* the "special" new robot that could think like a human had two brains that worked partially in tandem with one another, yet in other instances partly in opposition.) Why does our moral compass seem to be locked within the front part of the brain which scientists have so succinctly and pithily called the ventromedial prefrontal cortex (i.e. the VMPC)? (Oh yes, and why can't scientists come up with shorter names when classifying things? – but that is yet another even longer story.) As society has increasingly come to grips with the truth about human nature, science has flourished. This, in turn, gives us even more insight to our potential destiny as a species. The "ascent of humans" has now become quite literally the climb to space.

The fundamental question of the 21st century is whether humans will continue the "Bronowski ascent" and finally climb out of the Earth's gravity well to build permanent space colonies – or not? Ironically, as humans stand on the horizon of a great potential to leap forward into space, we also stand on the brink of potential destruction. These risks, as prompted by a burgeoning population and industrialization, include not only global warming and greenhouse gases, but also worries about the protective ozone layer that blankets the Earth and fouling of the oceans. We also have good reason to be concerned about industrial pollution of our biosphere, changes to the polar cap reflectivity, and even cosmic collisions with asteroids and other Near Earth Objects (NEOs). Such threats may not destroy the world or wipe out all life on earth, but such incidents could certainly threaten or destroy our advanced consumer-oriented global economy and eliminate our carbon-rich life styles. Today's teens take as "givens" that going off to the Mall, and other aspects of a rather luxuriant industrial civilization, are pre-ordained and constants. In fact, this life style is actually vulnerable in dozens of ways.

It is already known that there could be a specific and highly destructive potential collision with our planet. Our nemesis might be less than thirty years away. This would be in the form of a collision with a so-called

Near Earth Object (NEO) called Apophis. My long time colleague Apollo astronaut Rusty Schweickart, now with the Planetary Society, projects that this catastrophic collision could occur as soon as 2036. Although the odds are small, there is indeed the possibility that this good-sized chunk of real estate could crash into our planet and destroy much of our global infrastructure. Being crushed by an NEO is a scary prospect. As Sir Arthur Clarke has said, the reason that dinosaurs are not around today is they lacked a good space program. Humans may find that our space program did not measure up either.

Possible destruction of global civilization in 2036 is not a great time line. In fact this is a mere blink of the eye as measured in cosmic time. The extent of this threat and the likelihood of a direct impact are not clearly known today. But, what is known is that the Bush administration and its Office of Management and Budget, in an act of "far-sighted concern" in March 2007, moved to cancel NASA's plan to adequately fund a program to protect the planet against impact by meteorites, NEOs, asteroids, or comets. Apparently the urgent need to continue to fund an overseas war effort trumps planetary survival – at least for fiscal year 2008. As we know, the Iraq War has brought many tragedies, while other things, including this effort to protect our planet, have gone unsung and unnoticed.

Settlement on other planetary bodies, solar power satellites to beam clean energy to large scale antenna collectors on the Earth's surface, and permanent escape from Earth gravity are now seen by some as mankind's ultimate destiny. Yet, to most Americans and most people around the world, such ideas remain starry-eyed fantasy that can wait. The future, however, can wait only so long and then it disappears.

If we wait too long human history can go the way of Dodo birds and the T-Rex. In short we see the future of commercial space as being linked to ultimate human destiny. It is "vision" that makes humanity different, but most politicians are content to graze off conventional ideas, having little more vision than well-fed sheep. For every Al Gore willing to speak of an "inconvenient truth" there are hordes of Democrats and Republicans, as well as most political leaders around the world, who believe that visionary programs, commercial space and global warming are all things we can worry about tomorrow.

The ability to put the human genetic eggs into more than one cosmic basket is one way to strive to preserve the race. In order to understand the purpose and nature of life in the universe we must travel and explore by scientific probes beyond the bounds of Earth. Let's hope our destiny – at least one day – is to dump into the dustbin of history the now out of date saying: "The sky is the limit." Certainly the human exploratory limit now is at least the Solar System. Even though sub-orbital flight is still pretty pricey, it allows us to begin to think of a new tomorrow beyond the bounds of Earth.

The 21st century is a time where the purpose of some five million years of evolutionary history may become more clearly defined. Like the two faces of Janus, sophisticated technology offers both marvelous opportunities and serious concerns. Our "sophisticated lifestyles" are now characterized by high energy, high resource and high calorie consumption. When we link these lifestyles to an ever-burgeoning global population it begins to present us with some very nasty looking scenarios for our grandchildren. Massive floods, genetic anomalies that grow into monstrosities, starvation, and even worse could be the heritage we leave to our children's children. The so-called advanced industrial and technology-based society that we have been creating for many centuries is, in two words, both "fragile and perilous."

Coping with Future Compression!!!! The 21st Century Challenge

What does the emergence of space planes, space tourism, and citizen astronauts mean to the world and the future of *Homo Sapiens* as a race? One answer – and probably by far the most common among the general public – is that this is somewhere between "ho hum" and "small potatoes." To many, an hour or so jaunt on a space plane, in order to experience four minutes of weightlessness and a look at the big blue marble, is just the stunt of the decade. It is seen as no more than a passing fad to be replaced by sex on a hot air balloon or jetting around on levitated roller skates inside a spherical cosmodrome. Ironically, the Swedish government is indeed working on a new tax law that would put ballooning and personal space flights in the same tax category.

There is another possible answer – and this is actually quite a big deal. This answer suggests that private space commerce and parallel technological evolution heralds a whole new ball game for the people of Earth. This is a Future, however, that will be, at once, both exciting and dangerous. Technological progress can be seen as a one-way Stargate; we can go forward to the future, but not back again. Or more accurately, the only way back might be via another Stone Age. Technology presents us with many challenges. For instance when the first atomic bomb was exploded, the scientists were not sure whether such a blast might not ignite the entirety of the Earth's atmosphere. Fortunately global annihilation did not occur. The technology we invent, however, becomes more powerful and complex each year but also it tends to be more perilous as well.

We as a people have to worry about more than just global warming as a threat to human civilization – new technology offers us marvelous opportunities but reveals new threats to lifestyle characterized by the consumption of an expanding population. Yet, we may be on the threshold of wonderful new opportunities as well. Space planes and space tourism may eventually lead us to a wealth of new technologies. The future of commercial space is about far more than better rockets. New materials, space elevators, new cheap and clean energy sources, environmental solutions to global warming and much more could come from innovative new space systems. Commercial innovations in space may ultimately allow us to establish permanent colonies on the Moon and Mars. In time we might even seek to "terraform" Mars or perhaps even Venus or the Moon to create a new extraterrestrial biosphere where humans can live and breed a new generation of Martians, Venusians or Selenians. Even further in the future we might even explore beyond the Solar System.

Space commerce may thus reveal a future that could be an "upper," or it may be a "downer" – or it could be both. There is yawning before us a "Stargate" that promises a fundamentally different future for humanity. Space plane designer Burt Rutan claims we are on the threshold of a new tomorrow. "I know this is an interim step," he said. "Fifteen years from now, every kid will know he can go to orbit in his lifetime."[2]

The fundamental question addressed here is whether the 21st century is the time when the human race experiences fundamental change – a change that is not only likely but also perhaps almost inevitable. The timetable for mass transit to space may not be Burt Rutan's 15 years. It may even be 150 years, but this is a transition that humans must ultimately make if the species is to survive. The emergence of 'citizen astronauts' and the climb out of Earth's gravity is nevertheless humankind's ultimate destiny – the result of some five million years of evolutionary history. Let's review this history to understand why the 21st century apparently will mark a period of fundamental transition.

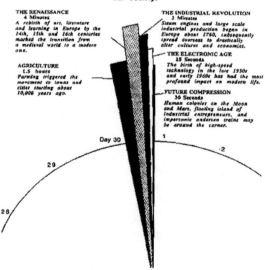

THE SUPERMONTH

A 30 day month illustrates the startling "Speed-up effect" of high technology in our society.

THE RENAISSANCE
4 Minutes
A rebirth of art, literature and learning in Europe by the 14th, 15th and 16th centuries marked the transition from a medieval world to a modern one.

AGRICULTURE
1.5 hours
Farming triggered the movement to towns and cities starting about 10,000 years ago.

THE INDUSTRIAL REVOLUTION
2 Minutes
Steam engines and large scale industrial production began in Europe about 1750, subsequently spread overseas to drastically alter cultures and economies.

THE ELECTRONIC AGE
15 Seconds
The birth of high-speed technology in the late 1950s and early 1960s has had the most profound impact on modern life.

FUTURE COMPRESSION
30 Seconds
Human colonies on the Moon and Mars, floating island of industrial entrepreneurs, and impactonic undersea trains may be around the corner.

Day 30

1
.2
29
28

Figure 10.1: Using the Super Month to Chart Human Development Over Time
(copyright J. Pelton)

First let's embark on a little time travel. Figure 10.1 below presents the "Super Month." Let's pretend that this picture is a kind of a basic time machine. Let's travel back some 5-million years ago to the time of the first Southern Ape men and women. Since that time, human civilization has changed in fundamental ways. The time of the evolutionary transition depicted in the Super Month is remarkably asymmetrical over the last five million years. In our Super Month we let each second represent 2 years. As can be seen in the graphic, it was 29 days and 22.5 hours into the Super Month – or 1.5 hours to midnight – when humans invented farming and went on to build permanent towns and settlements. Some four minutes to midnight represents the Renaissance, while two minutes to midnight is the Industrial Age. The last 20 seconds or so of Super Month time represents the time of television, computers, robots, space travel, lasers, cell phones, and spandex – all the great inventions of a high tech civilization. Generation X and Generation Next, at least in the "developed world," have difficulty in imagining a world that did not have these amenities readily available. These inventions and innovations of course represent only a flash in time across the millions of years of human existence. The 21st century appears to us today as a time where "FUTURE COMPRESSION" is reaching a point where the acceleration rate of invention is itself increasing – a phenomenon that physicists aptly describe as "JERK."

Maybe the SUPER MONTH analogy did not communicate clearly just how suddenly things are changing here on Planet Earth in terms of both the rate of change and the broadening scope of invention and the expansion of societal information. Here are some other examples that we hope helps to drive home the point.

- Imagine a building that is 10,000 stories or 20 miles high. If one wants to find when the Internet was invented in this building, as represented in the chronology of human history, it would be found a few inches from the ceiling of the top floor.
- Now, let's imagine this building in terms of stored global information or accessible data. Here we would find that the top 7,500 floors out of 10,000 stories would represent the information created since the start of the Internet.
- Finally, let's imagine a building with 150,000 stories that would extend more that 300 miles out into space. This would represent, assuming we keep the same scale, the stored information from human civilization we expect by 2025 if current trends continue.

The bottom line that the "Super Month" or the "Tower of Human Information" strikingly suggests is that we humans live in an increasingly interconnected, technology-driven world community where the "Future" continues to accelerate. The lives of those that live in this exploding technological society are ever more deeply impacted by "Future Compression." As the television commercial suggests: "Life comes at you fast." Tomorrow it will come at you much faster still. It would much easier for a Moses to come and live in the time of Thomas Jefferson and adapt quickly to a largely agrarian life style than for someone from the time of the Renaissance to come and live in today's world.

As this technological revolution is occurring, education, at least in the US, is falling behind, with only 52% of American children graduating from High School and in some cities like Detroit the graduation rate has fallen to one in four. US universities graduate more students from college in sports education than in engineering. Just four years ago 19% of the world's scientists and engineers were from India and China. Today that number is nearly 30% and the percentage continues to climb. US technological leadership, to the extent it exists today, is increasingly built on the skills of international immigrants.

If one looks at basic facts about the evolution of human society, the statistics about rates of change are more than a little alarming. Population growth, for instance, has been a worry for at least two centuries now. For a good number of reasons, ecologists to economists have begun to worry seriously about the so-called "Population Bomb." Since the time of ancient Greece and China the number of humans has grown from around 100-million to 6.5-billion or a stunning increase of about 65 times over the last 3000 years. During this time, the amount of information in society has grown from the equivalent of a few billion bytes to 20 petabytes (or 20,000,000 billion bytes). This means that stored information has expanded 20 million times or more than 300,000 times faster than human population. This is a very LARGE differential!!! This is a reasonably agile and aggressive snail making tracks to catch a space shuttle—and the differential is expanding.

Today we cope with information overload. Tomorrow it will be Mega-Information Overload. Commercial space innovations, is a part of the information explosion. If it is channeled in a constructive way, it could help us cope with the many problems that beset us from clean energy to global warming, from enough food to eat to improved health care and education services.

To Elon Musk, the future is about rockets that are hundreds to thousands of time better. But to Arthur C. Clarke, the great visionary of our times, commercial space is about survival and growth of the human species. In his vision rockets are merely "covered wagons" that will be phased out as truly advanced commercial space technologies evolve in coming centuries.

The Potential of Space Tourism: Entrepreneurial Talent and the Ultimate Frontier

Today, we may be on the threshold of wonderful new opportunities. We must hope and pray that space planes and the emergence of "space tourism" will be more than this decade's "hula hoop" fad. Space innovations – and in particular commercial space innovations that escape the confines of governmental space programs – could eventually lead us to a whole future for humankind. Who is to say right now exactly what this will be? It may lead us to new ways to cool the Earth's ice caps or monitor the fragile biosphere that allows humans to survive. If you visualize the Earth as an apple, the atmosphere that allows us to breathe and screen out deadly radiation is less than the skin of the apple. Human existence, as well as all fauna and flora that lives on our Planet, is and will continue to be a high-risk enterprise with such a fragile protective shield. What is this amazing new future? Is it Solar Power Satellites (SPS)? Or maybe it will be Extra-terrestrial relays to the Moon and Mars. Could it be the ability to mine asteroids? What about space elevators or satellites that can provide low cost distance education and health care? Or maybe it is something almost unimaginable today.

Maybe in a millennium or two we might even "terraform" a planet to create a biosphere where humans can live and breed a new generation of "extra-terrestrials" that look more like Dora the Explorer than something out of the War of the Worlds.

Some people, with a very limited vision of the problems besetting modern human civilization, may ask what is the purpose and value of having a space program? In fact, many even say the space program is a waste. They complain that space exploration and space expenditures take away valuable resources needed for education and health care. These people do not realize that space programs have allowed us to save tens of thousands of lives by predicting the paths of hurricanes, monsoons and typhoons. Our daily and weekly weather forecasts really start with weather satellites. The "Weather Channel" on television would be worthless without these eyes in the sky. Satellites have also saved stranded pilots and allowed communications to remote areas to provide relief to earthquake victims. Other satellites have allowed us to locate scarce resources from oil to iron as well as spot remote forest fires, diseased trees and blights on agricultural crops. Our servants in the sky help us follow rising ocean levels, spot tidal waves or volcanic eruptions, and to monitor the holes in the ozone layer. This layer protects us from the possible destruction of the species due to massive radiation-induced genetic mutation.

Two decades ago, in 1986 to be exact, as director of Project SHARE at Intelsat, the author helped to start the Chinese TV University. This satellite television educational project now has millions of students learning their lessons via satellite. India's Edusat program, that uses a similar model, now reaches nearly 30,000 villages and over a million students. The value of space programs today, in terms of lives saved, global communications, remote sensing, tele-education, tele-health, news distribution, and e-commerce is incalculable. Satellites, since they are out of sight, are thus out of mind. A Hughes Galaxy satellite failure that brought down the US paging system and cut off doctors from their patients a few years' back reminded people about critical satellite functions for a few days. But then they forgot again. Out of the news in the age of information overload means out of mind.

Arthur Clarke had his tongue only partially in cheek when he pointed out that dinosaurs did not survive because they failed to develop a space program. Our society is more dependent on space technology today than cavemen were dependent on fire. It is just that we don't know it. The question is still out. Are we humans, in a galactic sense, really any smarter than the departed dinosaurs. We may know the answer before the 21st century is out.

The truth is that our space programs are essential to saving the earth from destruction by comets, asteroids or other near earth objects (NEOs). Our "smart" satellites provide us with the tools needed to cope with the many hazards. Space programs have given us much more than Teflon, Tang and new kinds of plastics. Without weather and communications satellites, GPS navigation systems, remote sensing and surveillance devices modern society would be under-informed, our educational and health care systems weaker and human civilization would be much more at risk. The next time someone asks what is the good of space programs and satellites; you might respond that "survival of the species" is a pretty fair reason.

The thrust of most new space tourism ventures today is indeed thrill-oriented – a water flume in the sky if you will. Most ventures are simply seeking to provide Earthbound humans with the unique experience of seeing the planet from outer space and experiencing weightlessness. But there are many – including your humble authors – who see these ventures as a prelude to something that is much more profound – something that will impact the future of *Homo Sapiens* as a species. Rachel Armstrong, in her book *"The Future of Space Tourism"* in 2001, foresaw the possibility that humans would actually start going into space in a serious way within three decades, and speculated on the ultimate impact of such extraterrestrial activity:

"...space travel and space habitation pose more than an architectural challenge: they constitute a political and social issue embodying the cultural aspirations of the human race. With effective international cooperation, it is possible that in the near future people will travel into orbit to go on holiday and, perhaps later, settle there. The way in which these social spaces are constructed will shape the future of the human race physically and mentally. In the absence of the 'natural' world, space architecture will become the source of selective pressures acting on the human body. According to Darwinian theory, these forces will gradually cause extraterrestrial humans to evolve in a way different from that of their relatives back on Earth."[3]

The future of space tourism and where it will lead, of course, is unknowable at this point. Many will dismiss such speculation as idle science fiction fantasy. Even in the early 1950s serious scientists were still discounting space as not something that serious people should address in books of science.

In 1977, the late Princeton Professor Gerard K. O'Neill in his book *The High Frontier* speculated as to why a large number of people might actually go into space to live on a permanent basis. He suggested a variety of reasons. These varied from a sense of adventure, fears such as those related to nuclear or biological war, global warming or overpopulation, to the more practical need to generate new sources of energy from the sun. Other reasons ranged from a desire to perpetuate the species after a direct hit by an asteroid. Whatever the ultimate outcome, the first steps toward serious space tourism are now afoot. These efforts not only include developing the new technology and creating the business systems to make this a viable enterprise, but also the need to provide a regulatory structure to provide for the safety of crew and passengers.[4]

Let us again cite Sir Arthur Clarke – better known as Arthur C. Clarke, the noted scientist and science fiction writer who "invented" communications satellites in 1945 and went on to make (with Stanley Kubrick) the award-winning movie "2001 – a Space Odyssey" in the 1960s. He predicted many of these now unfolding events decades ago. In one of his "Egograms," written from his home in Sri Lanka, just a few months before his death in March 2008, the 90-year-old Clarke said:

> "Notwithstanding the remarkable accomplishments during the past 50 years, I believe that the Golden Age of space travel is still ahead of us. Before the current decade is out, fee-paying passengers will be experiencing sub-orbital flights aboard privately funded passenger vehicles, built by a new generation of engineer-entrepreneurs with an unstoppable passion for space (I'm hoping I could still make such a journey myself). And over the next 50 years, thousands of people will gain access to the orbital realm – and then, to the Moon and beyond. ... I have followed with particular interest the emergence of this new breed of 'Citizen Astronauts' and private space enterprise."[5]

Figure 10.2: Sir Arthur Clarke,
(Photo by Rohan de Silva, courtesy of Arthur C Clarke Estate)

Alas Arthur's dreams of his own space travels are not to be, but those dreams are very much alive in the legacy of his books and writings (the "Clarkives") and through continuing the efforts of organizations such as the Arthur C. Clarke Foundation.

Over the past fifty years, since the launch of Sputnik took the world by surprise in October 1957, there have been remarkable achievements to prepare the way for the era of citizen astronauts. The human adventure began with Yuri Gagarin, who made the first-ever manned space flight on April 12, 1961, completing one orbit of the Earth before landing safely in the Soviet Union. Three weeks later, on May 5, US astronaut Alan Shepard made the first sub-orbital flight by an American, to an altitude of 116 miles, and the race was on!

On May 25, 1961, President John F. Kennedy addressed the US Congress and provided the impetus and the funding in his famous speech:

> *"I believe that this nation should commit itself to achieving the goal, before this decade is out, of landing a man on the Moon and returning him safely to the Earth. No single space project in this period will be more impressive to mankind, or more important for the long-range exploration of space; and none will be so difficult or expensive to accomplish."*[6]

Thus began the story of the successes and the failures. Astronauts died in tests, but also walked on the Moon. Along the way came the end of the Cold War and eventually cooperation in space between America and Russia which led to the International Space Station. Today, space competition today is fueled by corporate innovation and the hope of building successful new space industries.

But over the years, governmental space programs have produced some impressive results. The Hubble and Chandra Telescopes have discovered new secrets of the Universe at its most remote reaches. The International Space Station has now been largely deployed and new rockets are being built to allow longer-term exploration of space. For decades, the advancement in space technology and space exploration has commanded huge budgets figured in the billions of dollars. These activities have been spearheaded by governmental space agencies such as NASA in the United States, ESA in Europe, the French Space Agency (CNES), the German Space Agency (DLR), the Russian Space Agency (Roskosmos), the Japanese Space Exploration Agency (JAXA), the Chinese National Space Agency (CNSA), the Indian Space Research Organization (ISRO), the Brazil Space Agency (INPE) and others. The US Space Shuttle and the Russian Soyuz programs have demonstrated that human access to space is possible with some regularity, and now much wider access to space appears to be fast approaching with China now set to play a key role in human space exploration, along with Europe and Japan as well. But, we are now ready for a new chapter to be written by commercial space enterprises. One can at least hope that these programs will see lower cost systems fueled by corporate innovation and entirely new technologies.

The most significant development in space activities in many decades is the exciting emergence of private enterprise as an important stimulant to the growth of new space businesses. As we have seen in earlier chapters, it is being led by a number of high profile billionaire entrepreneurs who are now bankrolling many of these new commercial space businesses. These billionaire backers of the new space commercialization industry, and others of their creative ilk, have become trailblazers in this remarkable new development.

The current evolution of this "space tourism" business, or more broadly the "space development" business, is in a state of rapid flux. The International Space Development Conference, held in Los Angeles, California in early May 2006, demonstrated the growing interest, not only by technologists and space enthusiasts but financial backers as well. Well over 40 private companies around the world are now embarked on this enterprise, and it is estimated that well over $1-billion dollars have already been invested in spaceports and new vehicle development programs to achieve sub-orbital and orbital flight with human crews and passengers aboard.

A little over a decade ago the whole idea of space tourism and reliable space plane operations on a regular schedule for booked passengers still seemed in the realm of science fiction. But as can be seen in the chronology provided in Appendix C, people have been talking about space tourism for nearly fifty years, although serious technical study and prototype development is a fairly recent event. It was the Ansari X-Prize competition and the success of the SpaceShipOne flight in 2004 that changed the landscape. At the same time, serious market studies by the European Space Agency, Futron and others are now projecting that space tourism and space commerce will become a multi-billion dollar enterprise.

However, the processes for international safety regulation, licensing and control are still being determined. And we believe that effective control of these processes, to ensure the safety of "space tourism" as well as the general public that might live near the burgeoning number of private spaceports around the world, is best addressed now and before a major commercial space accident occurs.

Widespread public involvement in space, plus active coverage of space exploration by journalists and media, can impact public perception. Creative new space enterprises and the development of industries that depend on access to space could be a so-called "transforming event" for both governmental entities, such as NASA, and commercial space ventures alike. Many of the forty-plus enterprises engaged in trying to design and build space planes or other vehicles to fly sub-orbital or even low earth orbit missions have already invested hundreds of millions of dollars in these new commercial ventures. In addition, hundreds of millions more dollars have been invested in the development and equipping of spaceport facilities around the world to support these new businesses once they become operational. This is clearly only the first stage of a new and emerging industry. While many of these enterprises appear likely to succeed and begin commercial operations within the next two to three years, a number of these will inevitably fail.

Many of the traditional aerospace companies as well as the space agencies still think they are in the rocket building business. This is a big mistake. The Conestoga Covered Wagon Company thought its mission was to build covered wagons rather than state-of-the-art transportation. The mission of today's space enterprises should be to design and build better space transportation systems. The goal of these companies and space agencies should be to provide transport that is more cost efficient, more environmentally friendly, and certainly a lot less dangerous. Chemical rocket propulsion systems will go the way of Conestoga Wagons – and within a few decades. They will be replaced by systems that are "smarter" and "greener" than the controlled explosion chambers we call rockets and space planes.

An important factor in the evolution of private space flight has been the stimuli created by several new and innovative space technology competitions, starting with the $10-million dollar Ansari X-Prize claimed by Burt Rutan and Paul Allen in 2004. The X-Prize founder Peter Diamandis, who many call the ultimate space entrepreneur (and who also helped to found the International Space University in 1983-1985), has long contended that private enterprise, stripped of governmental interference and involvement, can advance the development of a new and dynamic space industry if only the right stimulus is present.

The "prize" as a way to spur development in space transport has worked exceedingly well. This 21st century approach to commercial space has followed in the tradition of the American aviation industry where, in the early part of the 20th century, a $25,000 prize was created for the first aviator to fly solo from New York City to Paris. Charles Lindbergh, of course, was able to claim this prize shortly after World War 1.

This remarkably well-publicized and globally lauded event led to several more aviation-based prizes that stimulated innovation in the field of aviation and aerospace. The Guggenheim family, through a series of prizes and other initiatives including the founding of an aviation company, did a great deal to advance the cause of airplane safety and to create an airline industry in the US Now, in the 21st century, hundreds of millions of dollars from the commercial and entrepreneurial world were invested to try to claim the $10-million Ansari X-Prize. Paul Allen alone reportedly invested perhaps as much as one hundred million dollars in the SpaceShipOne development. Without Allen's backing Burt Rutan and his Scaled Composites crew would never have been able to claim the prize.

And so the momentum for commercial space development has accelerated since 2004. But it has not all been smooth sailing. As noted more fully in previous chapters, there are at least three elements that could be major problems for the evolving space tourism and space plane industry – perhaps ultimately even become "show stoppers." Clearly new and innovative solutions are needed for all three of these problems.

1. Environmental concerns. A number of environmental scientists are concerned about the potentially dangerous impact on the Earth's Ozone layer due to sustained and high volume flights to the stratosphere and beyond. There are also concerns about greenhouse gasses, ultra-violet radiation, "killer electrons" related to distortions in the Earth's magnetic field, solar storms and space weather. A low pollution, safer, lower in cost and essentially better way to access space is needed. A hydrogen fueled spaceplane engine might be a good place to start, even though water vapor at high altitudes also has its negative impacts. Better radiation shielding also is highly desirable.
2. The increasing trend toward the weaponization of space. This could be a problem in several senses. One is the possible future restriction of the private sector's current ability to fly freely into space due to space planes being potentially seen as "space weapons." A parallel concern would be increased apprehension among prospective space tourism passengers if they knew that their flight might be potentially seen as a hostile act by a country equipped by anti-aircraft missiles. Further ITAR and

other strategic arms restrictions might slow the development of new spaceplane systems. Today's space tourism flights essentially go almost straight up, arc over, and then back down. Such a flight trajectory may not trigger such concerns today. But, as capabilities mature and orbital access capabilities evolve, all of these concerns and more will undoubtedly arise.

3. Orbital space debris particularly in Low Earth Orbit (LEO). The proliferation of space debris poses an increasing threat to scientific, applications, and military mission satellites and to a lesser extent space planes.

Today it seems possible to meet and successfully answer all of these challenges. But ignoring these challenges as well as safety concerns, would be unwise. Here creative and active governmental and corporate interaction and cooperation is highly advised.

What is certain is that it will still be some time before commercial break-even (or profitability) is achieved by any of the players in the high-risk commercial space industry. And clearly one, and certainly two, accidents could, of course, scuttle the entire enterprise – possibly for them all. Ignoring the above issues could also prove dangerous and ill advised, not only for the evolving industry but to governmental oversight agencies as well.[7]

But the potential for break-through success is there for not only the commercial operators of space planes and space tourism businesses, but also for those who are looking beyond. Those that look to the longer-term future see not what is possible today but what could be. Visionaries such Robert Bigelow and the leadership of Inter Orbital Systems see that we could fly passengers not just into sub-orbital space, but to actual Earth orbit where they could stay in space hotels. Today, these visions seem more and more close to reality. Bigelow has his Genesis I and II inflatable "Transhab prototypes" for a future space hotel already in orbit. Inter Orbital Systems (IOS), of Mojave, California, has accepted a "deposit fee" from its first space hotel guest. This deposit will allow the near term launch of a scale model of their one and half stage to orbit launch vehicle-cum-space station.

Some other visionaries seek to deploy new types of space systems that will provide transportation that is much safer and lower in cost than today's space planes. There are scientists and engineers who believe there are cleaner and better technologies to put people into space than "controlled bombs." Proponents of the space elevator suggest that nano-tube carbon-cables that lift people into space would be a better way to access the heavens than lighting fireballs under passengers and pilots. Others see the need for a different business model for a new private space commerce business.

Some, for instance, envision that the technology being developed to sustain a space tourism business, may, in fact, also be key to developing a new business to build commercial supersonic jets that can fly from New York to Los Angeles at altitudes near 60,000 feet in about two hours. Such a new generation of SSTs would presumably ferry about the highest paid executives in the world and at enormous speeds. Already Lockheed Martin Skunkworks has designed the "QSS" as a prototype for such executive jet service. This supersonic business jet would carry 12 passengers at speed just a bit faster than Mach 2. Lockheed claims that for $2.5 billion they could bring such a "commercial executive space plane" to market. Instead of being a high cost flight to nowhere, as the sub-orbital space plane flights would be, this project would be definitely a high cost flight to somewhere. The European Space Agency has also financed research by Reaction Engines to examine the so-called A2 that could fly much larger passenger numbers from Europe to Australia at speed up to Mach 5.

These ventures, however, would still have the same potential show stopper concerns as the sub-orbital flights via space planes. This would be the potential environmental damage to the ozone layer that high altitude flights particularly pose.[8]

There is a cadre of far-sighted individuals who see beyond the space tourism business to a new tomorrow. They comprehend the potential of space elevators, space tethers, solar power satellites, low "g" material processing units, and a host of new applications that can make our world safer and better. They envision a whole new space economy. Some projects may even enable new technology and systems that can even rescue our planet from pollution and global warming. At least we can dream and hope for a better future.

As the song goes, Space Tourism "could be the start of something big." We just don't know what all these big things are yet. Let's hope we all get to stay around long enough to find out. Yet we do have some insights of what the future might hold.

Longer Range Efforts To Provide Private Access To Space

It is clear that private commercial endeavors want to launch people, cargo and satellites into orbit – even low earth orbit – in a more cost-effective and safe manner. But to achieve this objective, at least in the long run, will likely mean moving beyond rocket technology. The use of explosive chemicals is not only inherently dangerous but, from a longer-term perspective, it is also environmentally unsound.

In order to move beyond human's currently very tenuous access to low earth orbit and attain a reliable and sustainable presence in space, the costs of lifting payloads into Earth orbit must first be greatly reduced from their current levels. Equally important is the goal of increasing safety as well.

A number of concepts for lower cost, reusable launch vehicles have been proposed, but the simple and unforgiving nature of the equations relating to rocket propulsion underlines the limitations of presently known chemical propellants. Truly enhanced performance, cost reductions and enhanced safety via chemical rocket technologies alone seem a remote prospect. New technologies and rocket propulsion systems, such as metallic hydrogen, remain under study. But new fuels and chemical propulsion, even if achieved, do not hold great promise. It is not clear whether increased propulsion capabilities will lead to reduced costs or safer launch systems. But entirely new ideas are fermenting in laboratories. New concepts are stirring.

The following concepts about innovative ways to access space rely on new technology and new materials that allow alternative ways to escape the bonds of Earth's gravity.

Space Elevators and Tethers

A number of scientists and engineers, such as Dr. Brad Edwards, believe that it will be possible in coming years to devise a new type of system capable of carrying cargo and humans high into space on a ribbon of carbon thinner than paper. The robotic climber system needed to deploy these ribbons to the sky would be powered by highly concentrated laser beams or perhaps high performance solar cells that capture energy from the sun.

For decades, the space elevator, or the two-way space funicular, has been discussed as a concept by science fiction writers before being taken up by scientists and engineers. Arthur C. Clarke dramatically presented this idea in the book *Foundations of Paradise*. When Clarke penned this tale, it was seen as science fiction. As we know from Arthur's 3rd law what seem magic today can become tomorrow's reality.

Today, truly serious efforts are underway to develop the needed technologies to make such a space elevator system possible. This means intensive efforts to develop the materials (such as carbon based nano-tubes) and laser-powered (or solar cell enabled) climber robots. These technological developments are the most critical elements that might actually make such this seemingly fantastic idea operationally feasible. This micro-thin ribbon of carbon reaching almost a fourth of the way to the Moon might weigh as much as 14 tons. Although this might seem quite heavy it is, in fact, still amazingly lightweight when considered on a weight per kilometer basis. When it is worked out that entire ribbon would need to stretch some 100,000 kilometers (or 62,000 miles) into space, the weight per kilometre works out to be measured in hundreds of grams per kilometre or tens of ounces per mile.[9]

At the October 2006 X-Prize competition, NASA sponsored some $400,000 in prize money as part of the Centennial Challenges to stimulate the innovative concepts needed to develop a solar power robotic climber. Although this competition was not successful, the prize money has been increased for even more ambitious goals for future competitions. The NASA Tether Challenge calls for entrants to design and create super-strong, lightweight materials whose strength will be tested by opposing teams in a kind of tug-of-war competition.

Rather than dreaming up a shaft for a space elevator, the Tether Challenge is intended to encourage entirely new concepts in materials science. The Climber Challenge also requires teams to develop robotic "climbers" that lift the heaviest possible load, at a speed of at least a meter per second, to the top of a 67-meter (200 foot) "vertical racetrack" made of metallic ribbon. However, 67-meters is a far cry from the 35,580-kilometers needed to reach from the Earth's surface to GEO orbit, much less the additional

65,000-kilometers of tether needed to be deployed above the GEO orbit, in effect, to create the needed "negative gravity" above GEO orbit.

Okay, we admit "negative gravity" may seem more than a bit abstruse if not way over on the "weird" scale. So we will provide a little explanation about how the "space elevator" physics works. Arthur C. Clarke was among the first to explain the many uses that might be made of a rather magical geosynchronous orbit that encircles Earth above the equator, some 22,300 miles (or 35,800 kilometers) about the ground.

This special orbit, one tenth of the way to the Moon, allows a satellite to revolve around the world at the exact speed that the Earth rotates on its axis. Thus a satellite in the special orbit has the exact velocity needed to both stay in a circular path around the globe and complete exactly one revolution every day as seen from the perspective of the Sun. In this special orbit, where the pull of gravity is 1/50th of that at the Earth's surface, a satellite appears to exactly hover over a point above the Equator. (For the physicists out there who are keen to do some calculations, the pull of gravity at Geo is 0.22 m/sec^2. In contrast the pull of gravity at sea level on the Earth is 9.8 m/sec^2.) The ribbon that would connect the earth to a satellite in geo orbit, even if super thin, still weighs a lot and would pull the satellite down from its perfectly circular orbit. But if one also extended a ribbon outward from the satellite above the "Clarke" or geo orbit, the mass that is deployed above the satellite would serve to pull it back upwards toward a higher orbit. Calculations undertaken by Dr. Bradley Edwards of Carbon Designs, Inc., concluded that if one deployed a ribbon down from a geo orbit some 35, 870-kilometers, and also extended a cable outward for another 65,000- kilometres, the gravitation forces on the Clarke orbit satellite would more or less exactly balance. *Voila*, you would suddenly have a perfectly balanced virtual "elevator shaft" whereby it would be possible to lift mass to geo orbit. You could, of course, also use this shaft to lower mass down.

With such a space elevator you could also lift enough mass to geo orbit to assemble a vehicle to go to the Moon or Mars. Since the pull of gravity is so small at this altitude you could get to the Moon or Mars without a great deal of additional thrust. Ion engines or smaller scale rockets would be all that one would need once you had reached this very high altitude. In terms of escaping Earth's gravitation field this would be something like doing the equivalent of climbing to the observation deck of the Empire State Building and finding that to complete your climb you only had to ascend another few feet to the top of the radio tower from that already high point. Your mission would be more than 85% to 90% achieved.

One of the key firms developing the new carbon-based material technology that might have sufficient tensile strength to create a tether sufficiently strong enough is the Dallas-based start-up called Carbon Designs, Inc., headed by Bradley C. Edwards. Dr. Edwards is a former Los Alamos physicist and author of a book titled "The Space Elevator: a Revolutionary Earth-to-Space Transportation System." He is the advocate of constructing a ribbon made of new materials called carbon nano-tubes.

Nano-tubes are linked atoms of carbon in a unique spherical molecular form shaped rather uncannily like volleyballs. Scientists call them "Buckyballs" or "Fullerenes." (These names are a tribute to Buckminster Fuller. Inventor and materials guru Fuller correctly anticipated that this type of crystalline-like, high-tensile strength carbon-synthesized material could eventually duplicate and even exceed the tensile strength of diamonds.) Today's nano-tubes have a diameter 10,000 times smaller than a human hair. Thus, practical tether systems remain largely in the future and are more like laboratory experiments than industrial products. Yet nano-tubes of microscopic lengths are already being produced. Both their tensile strength and their electrical properties have attracted the interest of companies as small as Bradley's new firm, Carbon Design Inc., as well as industrial giants like IBM and Intel.

In the relatively near future, Bradley believes, Carbon Design Inc. will develop the first practical applications and on a large scale. These new solid, lightweight materials will be at least 10 times stronger than steel. These might be applied to the manufacture of golf clubs, tennis rackets or bicycle frames. This could then be followed for practical use in space within five to ten years, including the development of super light telecommunications antennas, and solar concentrators.

According to the Brad Edwards vision, the deployment would unfold in the following fashion. First, a single ribbon of nano-tubes – about a meter wide and barely thicker than Saran wrap – would be anchored aboard a huge ocean-going ship. This micro-strip ribbon would then be unrolled from spools, and taken aloft by rocket or lighter-than-air systems and ion engines into geosynchronous orbit, or to a height of 35, 870-kilometers (or 22,230-miles) high. Robot "climber" vehicles, powered by laser-like beams of light that relay electricity from solar panels aboard the mother ship, would then move up the ribbon, carrying fresh spools of nano-tube ribbon up to the top making the system stronger and stronger.

From there, smaller rockets would carry still more lengths of ribbon to a final point 62,500-miles up. At that point a massive counterweight would hold the entire ribbon in place as the earth's swift rotation keeps it taut – much the way a rock at the end of a string stays taut when a kid whirls it around and around. The elevator could be used by relays of "climbers" carrying entire spacecraft and supplies – even with astronauts aboard. A space elevator could be used as departure point for launch into deep space or simply in order to go on to the Moon or Mars. When you reached the top of the ribbon you are pretty much well on your way to anywhere. Advance ion engines or solar sails can take you from that point to pretty much anywhere you would want to go and at not very great cost or without the use of very much energy.

The NASA Challenge competitions for the best tether and the best "climber" systems are thus seen as a prototype for a new type of launch system for the future. This "space elevator" is envisioned as a flexible, cost effective, and hopefully much less risky launch system than today's chemical rockets that represent a highly controlled explosion. Estimates as to the timing, cost and safety of such a future system vary widely. The feasibility of deploying such a system, for a cost in the range of say $15-billion in a time frame of about 15 years in the future – at least this is Dr. Brad Edward's estimate– may be possible. This safe, cost-effective and highly flexible space infrastructure thus might be deployed at a cost that is on the order of 10% to 15% of that of the International Space Station (ISS) and possibly in significantly less time.

There are still many challenges to overcome and some skeptics feel that the Achilles Heel of this type of space ascension system is the heavy bombardment that the ribbon of nano-tubes would receive from the Van Allen belts that encircle the Earth. Super accelerated sub-atomic particles can do a lot of damage. Some fear the intense radiation and racing micro-particles could over time make Swiss cheese out of a thin ribbon of carbon hung from the sky.

This intense radiation might ultimately prove to be too severe for the space elevator to be reliably built. Yet, in terms of a space ascension system that would be safe, environmentally friendly and ultimately cost effective, this system could offer enormous potential if the scientists and engineers actually make it work.

Space Elevators: Why? When? And How Much?

Dr. Brad Edwards believes that NASA's current estimates that a space elevator would take 30 years to design and build are too conservative. Edwards suggests that if done by private enterprise with adequate funding it could indeed be in operation in 15 years.[10]

Today's space industry thinks very much within the "box." This is a box that reads "chemical rocket propulsion is the only answer." The large aerospace companies that design and build launch systems have nearly sixty years of experience of building chemical reaction rockets that do achieve their goal of putting payloads into space and with ever-increasing reliability. The Soyuz and Space Shuttle systems are on the order of 98.5% to 99% reliable. The designers of space planes are aiming at systems that are on the order of 999 out of a 1000 times reliable or better. The longer-term view of finding a more cost effective way to space with 99.999% reliability must truly think in terms of an entirely new technology. Rockets seem very unlikely to be able to achieve this kind of reliability. Add on the problem of pollution and cost and rocket launchers clearly do not spell a viable long-term way to go into space. The space elevator may be the best bet.

There are several reasons why a Space Elevator, given the development of the right new technology and the right operational systems and safety controls, could represent a quantum leap beyond either current or even future rocket technologies. The primary reasons are as follows:

- **Cost.** All rocket designs are governed by the rocket equation, which means that even the rockets of the future will weigh thousands of tons in order to lift a truck-sized payload to LEO, or a car-size payload to GEO. Only 5-10% of a rocket launcher system can be the actual payload. Even future nuclear propulsion will require on the order of a 10% payload to mass ratio.
- **Cost of Safety.** All rocket designs will have to burn tons of propellant *every second* in order to get going – such machines cannot be made cheaply, particularly if they are to be reliable, low risk and safe.

- **Ultimate Safety Concerns.** Rockets require an enormous continuous explosion. For this reason they cannot be made to be truly safe and 99.999% reliable. The huge amount of stored energy means that even a small problem in thousands of parts, chips, wiring, and subsystems can ultimately result in a catastrophic explosion. A Space Elevator climber can literally stop in its tracks if an anomaly is detected, and even slide back down if necessary.
- **Orbital and Deployment Flexibility.** The Space Elevator, as currently envisioned can offer flexible and large-scale deployment to LEO, GEO, and even Earth Escape trajectories.
- **Conservation of Energy and Environmental Concerns.** A rocket wastes most of its energy as propellant heat and velocity – only a fraction of the energy goes towards pushing the rocket forward. It also burns chemicals in the atmosphere and even more fragile stratosphere. A Space Elevator, on the other hand, derives most of its energy from the rotation of the Earth, and so needs to provide only a fraction of the energy – and even that, in the form of clean electricity, solar energy and laser transmission systems.
- **Reliability.** A Space Elevator climber, as currently envisioned, is simply a set of photovoltaic panels, laser transmitters, and electric motors. Even the rockets of the future will remain much more complex machines, continuously balancing fire and ice as close to an explosion as they dare push their designs.
- **Quality of Service.** Because of "gravity drag", rockets have to accelerate at 5-6 gravities, or they waste even more propellant against Earth's gravity. At such high forces, dynamic loads reach many tens of g's and the ride on top of the rocket becomes a very destructive experience. Most of the cost of building a payload goes into making sure it survives launch. A Space Elevator, on the other hand, transfers gently from 1 to 0 gravities, with no dynamic loads. The impact of this on the complexity, capabilities and cost of the payloads cannot be overstated.
- **Scalability.** The Space Elevator can grow in a linear fashion – if we double the thickness of the tether and add another power beaming station, the system can carry twice as much load. If we do this every two years for 20 years, the system will be able to carry 1000 times as much load. Rockets simply cannot increase their performance a 1000 times in 20 years even with innovative minds like Burt Rutan, Elon Musk or Richard Branson working on the problem.
- **No fairing size limitation.** Since it only moves through the atmosphere at slow speeds, the Space Elevator if strengthened enough over time could haul up an entire radio telescope dish in one piece – with no in-orbit deployment necessary, resulting in simpler payloads.
- **Quality of Service, Maintenance and Retrofit.** The Space Elevator can carry a communication satellite to GEO, test it in position, and bring it back down if it's not performing well. This sort of capability should also help to reduce insurance rates for space systems.
- **Reliability.** No sub-nominal launches. Once the climber reaches GEO altitude, the payload is in GEO – period. No circularisation burn will be necessary.
- **Up "AND DOWN" capabilities.** The Space Elevator is the only system capable of getting back down to Earth without the very risky maneuver of aerobraking. This again represents a major safety factor. The most dangerous single element in today's manned space missions tends to be the thermal protection system and the extremely high heat associated with the de-orbiting process. In short a space elevator or tether system can help us address the increasingly difficult problem of orbital debris.

There is no doubt that rocket transportation systems – i.e. chemical, electrical ion and nuclear rocket propulsion – can get much better and cheaper than they are today. Nevertheless, rocket launchers have fundamental characteristics and inherent problems due their basic physics. The laws of nature and the rules of motion, gravitation and explosive propulsion cannot be avoided. In the long run, a real space-faring economy will need a quantum leap in capabilities. Chemical rocket systems will one day be *passé*. The giant aerospace companies and space agencies have yet to accept this inevitability.[11]

Space Tethers: A Shorter Term Answer?

The Tether Unlimited Inc. organization[12] has proposed the use of Momentum-Exchange / Electrodynamic-Reboost (MXER) Tethers to provide a way to assist launching vehicles, payloads and people into orbit. This launch assist capability can combine with reusable launch systems based on chemical rockets or other advanced technologies to achieve the order-of-magnitude reductions in launch costs needed for a viable space economy.

The Tether Launch Assist concept involves a satellite-suspended tether that is permanently swinging through space that would act as a catapult to catch and then lift a first stage rocket launch to a low earth orbit. This rather unconventional approach would combine the technique used by Tarzan to swing through the jungle with the principles of a simple electric motor. This would create in space a system that would be capable of repeatedly picking up payloads from a sub-orbital trajectory (as synchronized to the pendulum swing) and then boosting the payload up to higher orbits. In this concept, illustrated in the figure above, a long and high-strength tether would be deployed from an orbiting facility. Then the tethered system would be set to swinging back and forth so that, at the bottom of its swing, the tip of the tether is moving much slower than the center of mass of the system.

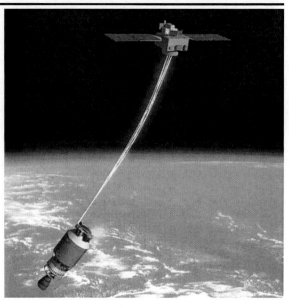

Figure 10.3: The Tether Catapult System for Raising Payloads to a Higher Orbit
(Courtesy of Tether Unlimited Inc.)

In this type of "dynamically stable" condition that would last for a few seconds a grapple system could be attached to the tip of the tether (or perhaps more likely an electromagnetic system). The idea would thus be to reach down below the facility and rendezvous with a payload moving in a slower, sub-orbital trajectory. The grapple (or electromagnetic system) would then capture the payload and pull it into orbit along with the tether system. It would then release the payload at the top of the swing, tossing it into a higher orbit. When the tether system captures and releases the payload, it transfers some of its momentum and energy to the payload. As a result, the tether facility's orbit will drop. To enable the system to boost additional payloads, the tether system can restore its orbit in between payload boost operations by using electrodynamic tether thrusting or perhaps restore its speed through the use of ion engines.

Electrodynamic thrusting would be accomplished by using onboard power supplies to drive current along the length of the tether. This current will interact with the Earth's magnetic field to generate so-called Lorentz JxB forces, much as the currents in an electric motor generate force. If the tether current is properly modulated as the tether swings through the Earth's magnetic field, these Lorentz forces will produce a net thrust that will restore the tether's orbit. If this seems a bit too technical, trust us, it really does work or electric motors wouldn't work. Because the electrodynamic re-boost technique utilizes the mass of the Earth, coupled through its magnetic field, as its reaction mass, it does not require expenditure of propellant. This enables the tether system to repeatedly boost payloads into orbit without requiring re-supply of propellant. The more basic approach, based on today's technology, however, that would allow the earlier implementation of this system would be to use ion engines. There would nevertheless need to be a discharge mechanism because the movement of the tether through the Earth's magnetic field would generate electrical energy in any event.

The net benefit of the Tether Launch Assist is that it can significantly reduce the velocity and thrust capability required of a reusable launch vehicle to launch a payload to low earth orbit. Due to the exponential scaling behavior of the rocket equation, even small reductions in the velocity increase (or ΔV) that a reusable launch vehicle must provide, thus translates into large reductions in launch vehicle size. Alternatively, if the launch vehicle system size is held fixed, the Tether Launch Assist can greatly increase the payload capacity of the launcher – or some combination of these two factors. In addition to the various space agencies, this approach has also been addressed by several private entities, but Tethers Unlimited Inc. (TUI) has perhaps invested the most time and energy into this approach. In order to see a demonstration of the tether lifting concept described above and to understand the concept in a completely visual way one should view their web site page.[13]

Nuclear Propulsion, Ion Engines, Solar Sails and Other Means to Access Space

There are a number of other means that might be utilized to access space or in time to travel across the solar system. Unfortunately some of the most efficient means of travel around the solar system are really not viable ways to escape the Earth's gravity well. Some of the means that might be used to travel from GEO orbit to the Moon or Mars or other destinations such as ion engines, solar sails, etc. do not have sufficient propulsive force to lift off from the Earth's surface and escape the gravitation forces that exist at lower altitudes. At this stage the only other propulsive force available that might allow lift off from the Earth and then power sustained space travel is nuclear propulsion. NASA and other entities have development programs underway and these have become increasingly more viable over the past two decades.

Nuclear propulsion systems have the ability to overcome the specific Impulse (I) limitations of chemical rockets because the source of energy and the propellant are independent of each other. Nuclear propulsion energy for a space probe would typically come from a critical nuclear reactor in which neutrons split fissile isotopes, such as 92-U-235 (Uranium) or 94-Pu-239 (Plutonium). This splitting of atoms in turn releases energetic fission products, such as gamma rays, and enough extra neutrons to keep the reactor operating. The energy density of nuclear fuel is enormous. For example, 1 gram of fissile uranium has enough energy to provide approximately one megawatt (MW) of thermal power for a day. Thus one kilogram would produce the equivalent of one gigawatt (GW) or a billion watts of thermal power for 24 hours.

The heat energy released from the reactor can then be used to heat up a low-molecular weight propellant (such as hydrogen) and then accelerate it through a thermodynamic nozzle in same way that chemical rockets do. This is how so-called nuclear thermal rockets (NTR's) work.[14]

The US Los Alamos National Laboratory has performed basic studies on NTR beginning in 1955, and NASA also began it own studies in the late 1950s. The first ground test of an NTR, designated "KIWI-A", was conducted in 1959 by Los Alamos, leading to a formal NTR development program, designated "Nuclear Engine for Rocket Vehicle Application (NERVA)." NERVA was under the direction of NASA and the US Atomic Energy Commission.

Researchers at the Los Alamos National Laboratory subsequently have considered another interesting NTR propulsion scheme, shown in Figure 10.6 below, and known as a "gas core nuclear rocket (GCNR)." In a GCNR, hydrogen is pumped into one end of a cylindrical reaction chamber, with an exhaust at the other end. The hydrogen expands as it passes through the chamber, and not all of it goes out the exhaust, instead flowing back up the chamber. This creates a toroidal vortex of hydrogen gas that can be used for fission reaction containment. Dust-sized particles of uranium are injected into the toroid and accumulate at its center.

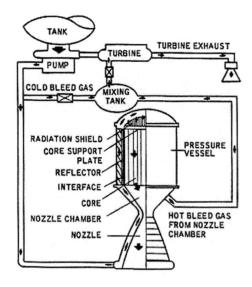

Figure 10.4: Schematic Diagram of a Solid Nuclear Thermal Rocket (NTR) Engine

An alternative approach to NTR's is to use the heat from a nuclear reactor to generate electrical power through a converter, and then use the electrical power to operate various types of electrical thrusters (such as ion engines) or by using a magneto-plasma-dynamic (MPD) approach. The MPD can operate using a wide variety of propellants (hydrogen, hydrazine, ammonia, argon, xenon, or even Fullerenes).

To convert the reactor heat into electricity, thermoelectric or thermo-ionic devices could be used, but these have low efficiencies and low power to weight ratios. The alternative is to use a thermodynamic cycle with either a liquid metal (sodium, potassium), or a gaseous (helium) working fluid. These thermodynamic cycles can achieve higher efficiencies and power to weight ratios. (See Figure 10.7 below). No matter what type of power converter is used, a heat rejection system is needed, meaning that simple radiators, heat pipes, or liquid-droplet radiators would be required to get rid of the waste heat. Unlike ground-based reactors, space reactors cannot dump the waste heat into a lake or into the air with cooling towers.

NEP (Nuclear Electric Powered) rockets are much more complex than NTP (Nuclear Thermal Powered) rockets, but it also produces better results. In NEP, heat from an on-board nuclear reactor is converted to electrical power. An electric thruster then accelerates ions or plasma to a very high velocity. The specific impulse of NEP is considerably higher than that of NTP. The Deep Space 1 spacecraft, powered by a ion engine, for example, has a thrust of only 92 milli-Newtons. To reach a useful velocity, however, NEP engines must operate continuously for years, while NTP engines need to operate for only an hour or so. In short NTP engines might be used for Earth lift off while NEP powered ion engines might be used for travel within the solar system.

Both the United States and Russia, in particular, have actively pursued research and development in space nuclear power and propulsion. In the United States, radioisotope thermoelectric generators (RTG's) were developed in the SNAP program (Space Nuclear Auxiliary Power) and used in early test satellites in the 1960s, the lunar landing missions of the early 1970s, and most planetary space missions since then (Pioneer, Viking, Galileo, Ulysses, and Cassini).

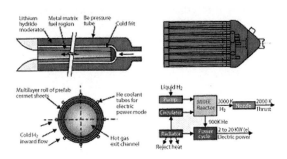

Figure 10.5: Nuclear Thermal Propelled (NTP) rocket engine

The most advanced concept in nuclear propulsion remains nuclear fusion systems but this technology currently appears decades away for practical applications to space systems, while RTG, NEP and NTP systems represent a range of capabilities to systems that are flight ready today to systems that are a decade away from practical use.[15]

There are a number of problems with nuclear propulsion that go beyond the realm of physics and engineering. One obvious issue is that of environmental risk. Several nuclear powered satellites have come back to Earth in an uncontrolled manner and as a result created nuclear release radioactive hazards. Secondly, because nuclear materials are closely linked to weaponry with the gravest danger being the creation of so-called "dirty bombs," these programs are overseen as classified governmental projects. This obviously limits private initiatives and entrepreneurial innovation. Ion engines, however, are not so constrained and offer considerable opportunities for private sector innovation.

Various forms of ion engines beyond nuclear powered systems are the subjects of intensive study. These include such ideas as using solar concentrators and high efficiency solar sails to provide electric power to high performance ion engines that might be effectively used for trips to Mars. There is also a wide range of studies underway to develop new materials that could be used to build solar sails that could propel spacecraft around the solar system as well. Many of these propulsion systems also envision the use of gravity gradient acceleration that has spacecraft falling near the Sun and then using a slingshot effect of solar gravity to attain much higher speeds. There are yet other ideas such as levitated mass drivers that would work most efficiently on the Moon.

The point of these various possible technologies for 21st century space access and travel is that none of them are yet mature, but research might make them viable within the next decade or two. Almost all of the concepts whether it be space elevators, space tethers, nuclear propulsion, ion engines, solar sails, solar electric powered ion engines, gravity gradient acceleration or levitated mass drivers, are likely to prove preferable, at least in time, to chemically powered rockets. This is because chemically fired rockets have many flaws that include:

- Problem of uncontrolled explosion and thus with overall astronaut safety
- Thermal safety issues
- Inefficient total mass to thrust ratios
- High cost

- Environmental damage to the Earth's atmosphere via venting of greenhouse gases and destruction to the Ozone layer.

The twentieth century and the early part of the twenty-first century will be remembered as the age of chemically-fueled transport for the ground, the sea, the air and space. By the second half of the twenty-first century, however, new transportation systems and new propulsive technologies that are lower in cost, friendly to the atmosphere and safer will arrive. These new power and transport systems will lead to the phase out of the age of petrol-fuels and chemically powered internal combustion engines. Here the fields of aerospace engineering, electrical engineering, chemical and mechanical engineering may find new planes of cooperation to find critical new answers to continue human commerce on the ground and space research in the skies.

Impact of new technologies

New propulsion technologies, new solar power systems and concentrators, innovative fuel concepts, new materials, new breakthrough in electro-magnetics, nuclear and quantum physics, new types of escape systems, new thermal protection systems, new lighter than air systems, and new remote piloting techniques are just some of the R&D currently underway that could make a difference. These and other technologies could and should make space planes and space tourism safer, more reliable and less costly. Insurance and capital investment for space tourism flights will become more attainable. Space access will become not only lower in cost but much safer as well.

It is unclear how much new technology will evolve and how soon it can be effectively applied. Some suggest that the technology, in an openly competitive market will come more quickly and costs will fall sharply. It has even been suggested that rocket systems and space planes will follow the pattern of rapid development that occurred when computers shifted from mainframes to personal computers. "Just as the personal computer revolution dramatically increased performance and lowered the cost of computing, the market for personal space flight promises to transform the economics of space operations," said Gregg Maryniak, Executive Director of the X Prize Foundation in St. Louis, Missouri and spokesperson for the new Personal Spaceflight Federation.[16]

Only time will tell if we will see such a startling revolution in space flight. Hype hardly ever defines the future. Dedicated research and visionary thinking fueled by entrepreneurial management and investment can make a difference. If computer technology had been confined to a governmental program it seems highly doubtful that we would today have palm top devices with many thousands times the capability of the Univac computer that required thousands of radio tubes and a room full of electronic components.

Yet sometimes what seems hyperbole today turns out to be the insight that can create a new totally new business or lead to a breakthrough technology. Here's to the day when chemical rockets are consigned to air and space museums and conventional rocket launches can be viewed on I-Max screens. In more ways than one, today's space planes are a pathway to tomorrow.

In short the key to thinking about space planes and private commercial space is "innovation" and not "rocket technology." Today's space planes will be viewed in museums someday in the same way as we look at Conestoga Covered Wagons and the Univac computer — important gateways to new technology and radically different ways to view the future.

References:
1. Joseph Bronowski, *The Ascent of Man* – Little Brown & Co., London -August 1976
2. Cathy Booth Thomas, "The Space Cowboys" *Time*, March 5, 2007 p.55
3. Rachel Armstrong, "*The Future of Space Tourism,*" in John Zukowsky, editor, *2001: Building for Space Travel*, p. 175 – Harry N. Abrams Inc., 2001, New York
4. Gerard K. O'Neill and Freeman Dyson – "*The High Frontier: Human Colonies in Space*" – Apogee Books, New York, 1977.
5. Arthur C. Clarke, Egogram for December 2006, via e-mail distribution
6. NASA History Office – http://history.nasa.gov/moondec.html
7. Mary Evans, "Rocket Renaissance: The era of private spaceflight is about to dawn" – The Economist, May 11, 2006 -http://www.economist.com/science/displayStory.cfm?story_id=6911220
8. "All Sonic, No Boom" *Popular Mechanics*, March 2007, Vol. 270, No. 3, P. 64-67
9. David Perlman, "NASA Offers Prize for 'Space Elevator': Beams of Light Could Propel Cargo, Human" – *Chronicle*,

April 25, 2005.
http://www.sfgate.com/cgi-bin/article.cgi?f=/c/a/2005/04/25/MNGM4CEAPG1.DTL
10. Space Elevator 2010 – web site
www.elevator2010.org/site/competitionClimber2005.html
11. Carbon Design Inc. Web Site – http://www.carbondesignsinc.com/
12. Tethers Unlimited web site – http://www.tethers.com/LaunchAssist.html
13. Tethers Unlimited web site – http://work.offshore.ai/miguel/tethers/tetherl3.html
14. Blair P. Bromley, Space Exploration Technology – http://www.astrodigital.org/space/nuclear.html', 2000-2001'
15. Blair P. Bromley, Space Exploration Technology – http://www.astrodigital.org/space/nuclear.html', 2000-2001'
16. Leonard David, "Personal Space flight Leaders Eye New Federation," space.com, February 9, 2005 – http://www.space.com/news/personal_space flight_050209.htm

— Chapter 11 —

"Just as the personal computer revolution dramatically increased performance and lowered the cost of computing, the market for personal space flight promises to transform the economics of space operations." – Gregg Maryniak, The X-Prize Foundation.

THE TOP TEN THINGS TO KNOW ABOUT SPACE TOURISM AND THE FUTURE OF SPACE

So there will be more and more regular people – not super astronauts – flying "beyond the skies" on commercial vehicles and within the next two years or so. What does this mean for humanity and the future of the human race?

Most of you will have a ready first response. Big Deal!!! So, fairly soon a handful of fat cats and celebs can pay big bucks to fly for a few minutes into space. Well actually it is not "truly outer space" – just so-called sub-orbital space. The billionaires, movie idols and rock stars just can't find enough ways to flaunt their material wealth, can they? With greenhouse gases rising and people starving in Darfur, a $200,000 ride that lasts a couple of hours or so seems to some just a bit decadent.

Certainly this new 100-kilometer high club may get some high-g kicks and prove the Earth indeed is not flat. But won't this trip to nowhere just burn up a lot of extra fuel? Might not all this galactic joy riding into space lead to faster global warming? Couldn't all those rocket planes burn a bigger hole in the ozone layer to boot? In short, so what? Why should anyone who is serious about global warming, famines, better health care services or truly "serious issues" care if the sub-orbital space tourism business takes off or not?

Some of you may say okay. Let the joy riders "trip out" – literally and figuratively. Let all of us just enjoy the time we have left before humans flood the earth when the ice caps melt. Or before greenhouse gases destroy the biosphere or we otherwise screw up the planet so our great-grandchildren can't live here anymore. Most, however, are likely to ask what does all this have to do with me and my family and my friends down the block? Isn't all this expensive, high-end space tourism business pretty irrelevant to the average John and Jane Doe?

The man and woman on the street – let's say Mary in her condo or Ralph in his three bedroom rambler–are most likely indifferent to this outer space business. Mary and Ralph are likely to assume that these new commercial space programs are pretty irrelevant to them and their lives. They probably believe that sub-orbital space tourism has nothing to do with serious space exploration, saving the planet or the long-term destiny of humankind. Right? We actually think – wrong!!!!

First of all, the space tourism industry does need to be re-thought and its technology improved. But all of this will come in time. Likewise NASA and possibly some of the other space agencies around the world need to be re-invented to become more relevant, more cost conscious and better focused. We actually believe that the most effective and innovative space agency in the world may well be the Indian Space Research Organization. ISRO has had some remarkable achievements in the last few years and with a much lower budget that the big space agencies. NASA, in contrast, has been the most "success-challenged," particularly when judged in terms of "bang for the buck."

Barack Obama, who campaigned in the 2008 US election race on a platform of "Change," said he grew up on Star Trek and is awed by the potential of space, but he feels that NASA in recent years has "mostly lost its way."

Certainly, the FAA Office of Space Commercialization, that oversees commercial space and space tourism ventures needs to keep evolving and adjusting its role to new realities if it is to be a successful regulator of the new commercial space frontier. At the top of the list for space and aviation regulators are some new and creative ideas to keep space planes and high altitude supersonic business jets from doing damage to the Earth's atmosphere. It is critical to make sure that future flights of all types (conventional, supersonic or hypersonic) do not end up depleting the ozone layer or overloading the atmosphere with greenhouse gases. This fragile layer, now with a sizable hole in it at the Poles, is what protects humans from rampant genetic mutation and species extermination. Personally we feel that preservation of the human race is sort of a high priority—up there in importance and newsworthiness with the shenanigans of a Paris Hilton or Britney Spears. The following is why we believe that space commercialization is truly important to the longer term future of our great, great grandchildren.

We believe that commercial space entrepreneurs are the people that can and will find new solutions, help us develop clean and cheap ways to fly across the world and even into the heavens. The conventional space agencies are typically too large and bureaucratic to find the BIG SOLUTIONS needed to get humanity through the 21st Century. In short, we need to give a new team of start-ups and entrepreneurial wizards a chance to find totally new answers, new systems and new environmentally smart technologies. So what are we talking about?

We need to find smarter ways to get people into space than just putting controlled bombs underneath them. To launch a rocket today we are, in effect, lighting the fuse on a just controlled explosion. We need something like clean hydrogen fueled vehicles that can fly people around the world further, faster and cleaner. In time we need totally new systems like space tethers, space elevators, or advanced ion engines that can get people and "smart machines" into space in ways that are a lot cheaper and cleaner.

There are a lot of things we need to do smarter and better, but do them we must. Cheap and reliable access to space is our destiny – or our destiny is to become yet another failed species. Over the longer term, access to space that is safe and lower in cost is essential to maintaining a viable human civilization on Earth. What might be called "space tools" are key to combating global warming, detecting and taming hurricanes, warding off threats of comets, meteorites and near earth objects, developing better ways to cope with disasters, finding vital resources, improving agriculture and monitoring our oceans and coastlines, sustaining global communications networks and supporting vital infrastructure. A recent New York Times editorial noted that there are probably at least 1,100 meteorites and asteroids out there in Earth's neighborhood – i.e. Near Earth Objects (NEOs) – that could either destroy the planet or do a great deal of damage. The current US space program strategy, as defined from the White House, can currently be called the "Let's ignore it and maybe nothing will happen to us Program."[1] My good friend and space associate Apollo 9 Astronaut Rusty Schweickart is sufficiently worried about the destruction of modern global society by a huge meteorite crashing into Earth that he has formed "the B612 Society" in cooperation with the Planetary Society to create a systematic response to these dangers. In case this all seems way too theoretical, just be aware that there is a quite solid chunk of real estate out there projected to come perilously near to Earth in 2036.

But we are concerned about more than scary meteorites. We humans are a rather scary race. We could wipe ourselves out by nuclear accident, global pandemic, rampant mutation of the species if the protective ozone layer dissipates or by a slow roasting that comes with excess buildup of green house gases. We need a back up Plan B if our biosphere here on Earth should some how fail to sustain future generations. This may seem heavy stuff, and way too pessimistic; but a back up plan is a part of what commercial space is all about. Let's build a new type of basket in which to keep and protect some of our eggs.

Cheap and effective access to space is also needed for a lot of other good reasons. Seeking to "fly beyond the skies" is also required to discover the mysteries of our universe, tap into clean solar energy, monitor our oceans and atmosphere and much more. We may not succeed in a quest to preserve the species in two different cosmic locations in the next decade, or next century or even next millennium, but the challenge remains. The Leif Ericksons, the Columbuses, the Magellans, the Galileos, the Henry Cabots, and the Lewis and Clarks of the future just want a chance. We must not say no to the future. We should not say to explorers of the 21st and 22nd century – to our future progeny – that we lost our human curiosity. We no longer dare to believe there could be a better and more profound future for our great, great grandchildren.

We believe that it is part of the genetic code of *Homo Sapiens* to say that we cannot know enough. We are destined to explore – to go where no one has gone before. We can never abandon our own sense of imagination that has defined the human experience for millions of years. The human species will never survive if it suddenly says we are now content to stay home and abandon future voyages of discovery. We give up on the future.

Humans badly need a new stairway to the stars. If we don't find this new way to get into space within the next few generations we may be in deep trouble.

Commercial space business, if nothing else, is rich in innovation. This is why space tourism must be recognized as just the start of something big. Let's just hope we get it right – and in the not too distant future.

Private spaceflight is thus likely to be the beginning of a fundamental revolution in the feasibility, affordability and accessibility of space for a growing number of humans across the planet. The vision is that in time millions of humans will go into space. In short, commercial space travel must become much more than a lark.

There will be missteps. Mistakes have been and will continue to be made. But, in the long run, the history of humankind will ultimately be divided into two distinct parts. Phase one is when humans lived, breathed, bred and incubated within the confines of the Earth's gravity. Phase two will be when humans expanded the scope of their activities into the broader reaches of the Solar System and reached beyond the skies. Affordable access to space will ultimately be critical to the survival of the human species.

Figure 11.1: The Saturn V that Lifted the Apollo 11 Mission to the Moon Performed Faultlessly
(Photo Courtesy of NASA)

Neil Armstrong had it right when he said: "One small step for a man. One giant leap for Mankind." Well almost right. Perhaps he should have said "One giant leap for Humankind." Any new celestial civilization won't get very far without some women around to sustain the species and provide some much needed insight and common sense as well.

In 1969, at the time of the first Moon landing, our hopes were high and our expectations were even higher. Futurists predicted that low cost trips into space would be commonplace in a decade – twenty years at the most. Cal Tech scientists actually projected that astronauts would fly to Mars—in the 1980s!! Allen Drury wrote a popular novel about a trip to Mars in the early 1970s that he believed was only a short time in the future. Remarkable achievements in terms of projecting rapid and economical access to space by the Space Shuttle turned out to be wrong by orders of magnitude. Ever since Apollo, NASA's space program has had more than its share of troubles. The addition of a solid fuel rocket to the Shuttle was only one of many misjudgments.

Wernher von Braun, on the occasion of Apollo 17, the last Moon mission in 1972, predicted that there would be a "permanent Lunar Colony" and that it would be achieved by 2009 or 2010. When news reporters asked why his prediction was so precisely timed, he responded that it took us 37 years from the time we first reached the South Pole until the time we established a permanent colony in Antarctica. He calculated that the challenge of setting up permanent shop on the Moon, in light of ever-increasing technical know-how, should allow us to create a realistic colony on the Moon's surface in 37 years.

Well, we all know he was wrong, and off by at least a decade if not by two or three. One reason is that von Braun envisioned a space station as a stepping-stone to the Moon. This was a BIG mistake. It turns out that a large-scale space station was an unnecessary and burdensome step to the Moon and Mars. Clearly NASA and its partners made a series of colossal errors in the design, construction and operation of the International Space Station. Likewise NASA's mistakes involving the Space Shuttle design and its multi-billion dollar refurbishment program is almost legendary in terms of compounded errors. The

Challenger and Columbia disasters made all the headlines, but this was only the tip of the iceberg. The mistakes and errors with the Space Shuttle could fill a number of books. (We should know, since a number of these have already been written.)

But the ISS and the Space Shuttle are now the "back story." These astronomical mistakes are today, at least in terms of galactic history, largely space flotsam and jetsam over the cosmic dam. The point is that national space agencies are pretty good at developing totally new and state-of-the-art technology. The scientists and engineers at space agencies are very proficient at solving totally new and arcane problems. But, NASA and its counterparts around the world are generally "piss poor" (pardon our French) at implementing major new infrastructure projects against a budget and achieving a drop-dead schedule.

Figure 11.2: The International Space Station as of late 2006 – Over budget
and subject of debate in terms of its utilization and usefulness.
(Photo Courtesy of NASA)

Rather astonishingly, when Ronald Reagan first announced the "Space Station" in the early 1980s, it was supposed to be finished by 1994 and it was to cost a mere fraction of what the ISS will eventually cost. When and if finished in 2010 or 2011, the ISS will cost about $100-billion. This works out to about a billion dollars a ton. No commercial contractor could undertake a large-scale project and then finish it some 15 years late and with a price tag four or five times over budget.

If NASA were a business it would have been bankrupt and out of business years ago. Nor is the ISS the only example of some galactic errors of judgment and management. NASA has started and then abandoned at least six space plane projects at a total cost exceeding billions of dollars. The time has arrived for private enterprise to lend its hand to make space projects affordable and responsive to schedule.

We are indeed seeing a new type of "private space venture." These initiatives – often backed by cagey and creative billionaires–are seeking to develop commercial space plane technology as well as operate a profitable space tourism business. The business organization and commercial plans of these new commercial ventures were described in earlier chapters as the "anti-NASA."

These new, agile and entrepreneurial organizations for the most part totally reject the space agency model. They certainly do not have thousands of employees formed into large bureaucratic units supported by large aerospace behemoths with even larger teams of highly paid employees. These totally new types of private space organizations certainly seek to avoid operating under bureaucratic processes that can squelch creativity, prevent innovation, or frown on flexibility of operation. The creative "Lone Ranger" has a difficult time making it to a leadership position within NASA. The one-time "rebel" Pete Worden, who now runs NASA Ames, seems to be an exception. But even the free-wheeling and entrepreneurial Worden, who ran the shoe-string Clementine mission to the Moon while in the Air Force, now seems much more the conformist as he has risen to the top ranks of the US space agency.

The processes found within space agencies for bureaucratic procedures, group think, quality assurance, independent verification and validation are so cumbersome that a concern for safety can get lost in the process. Then after creating tight standards thousands of "waivers" to safety standards are then allowed. At times it does seem a lot like Alice and Wonderland at the need to believe without any problem, any number of impossible things before breakfast.

This fresh new private sector approach to space projects can, we hope, bring new economies, new applications and products, new opportunities, new technologies and new ways to address a variety of problems both in space and on earth.

At this stage, no one can say what our future will look like? Pelton's law of forecasting explains why one should never make a 5, 10 or 25 year prediction. Always go for a 50-year prediction when no one will be around to check up on your accuracy.

Current concepts for a spaceport waiting room (See Figure 5.1 in chapter 5 for example) may look amazingly retro in a few decades. The point is that most people imagine it as a Buck Rogers fantasy rather than as a true doorway into tomorrow. If we manage it right, these spaceports can be a gateway in both a literal and a conceptual sense. The best of commercial space technologies could help us find new sources of clean energy, combat carbon pollutants and even re-invent the global economy where the cost of planetary sustainability is priced into the cost of goods and services.

Peter Diamandis' X-Prize enterprise that started the whole spaceplane industry has already morphed the X-Prize franchise into a challenge process to develop better and safer ground transportation, better and cleaner energy systems, and much more. One of the more interesting new X-Prize challenges is the hundred miles per gallon automobile. Diamandis has said for many years that it is the pioneering individuals willing to think outside the box and challenge today's conventions that change the world. Committees and groups of people "defend the status quo."

The Big Questions

The previous chapters have set out the story to date about the commercialization of space and the coming new era of citizen astronauts. A number of billionaires, a gaggle of rocket scientists and many dozens of innovative design engineers are hard at work. These designers are working in facilities that range from sophisticated labs to back lot garages. All of these efforts are aimed at creating viable space tourism businesses – mostly in the US, but there are others in several different countries hard at work as well.

There remains a host of big questions, however, not yet answered about this new commercial space industry. When will safe, reliable and economical space planes and spaceports become operational? Will these efforts closely parallel the development of high altitude supersonic executive jets and will these two industries ultimately be closely interlinked or not? How will these various efforts be regulated at the national and international level? When will the cost for fare-paying passengers reduce to a level so that a significant number of people can be attracted to the experience of a flight in space? Will a competitive, commercial market lead to more opportunities to expand into a wide range of new space-related businesses – and how safe will these activities be? Can private space ventures lead to affordable stays in space in "private space hotels"? Can the private space industry lead us to clean and economically viable solar power satellite systems, or allow cost effective in-orbit systems that can manufacture better and cheaper pharmaceuticals, building materials, or entirely new products? Can space commercialization lead to clean new hydrogen rocket engines or other "green" ways to approach both air and space travel? One of our key challenges over the next two or three decades is to find ways that move people in much more environmentally friendly ways not only from Earth to Mars but from Cleveland to Milwaukee or from Singapore to Sydney. And, one may hope this is just the start.

We must hope for second generation breakthroughs as well. In time we could and should look to new ways to access space such as space elevators, space tethers, advanced ion thrusters, nuclear powered systems or solar sail systems? What will be the relationship between private space ventures and national space agencies as traditional roles change due to changing economics and opportunities? In short, where are all these private space flight ventures headed within the next decade and over the longer term? Will this changing role be different in the US and other countries? Will NASA and the FAA need to be restructured to adapt to the presence of serious commercial space activities? Where will ventures in Australia, Canada, China, France, Germany, Israel, Russia, Singapore, the United Arab Emirates, the United Kingdom and other countries fit into the new space tourism industry? Who will succeed and who will fail?

There have been some extravagant claims about dramatic changes that will come as the result of private space initiatives. But, it is simply too early to know if this is the hype one expects as a part of the marketing and PR strategy for these new companies – or not. What will these companies be saying in five or ten years' time? If they fail to get an acceptable return on their huge investments, it will likely be a much different story. Can space agencies adapt successfully to a new era in which it is private companies that provide an increasing percentage of the new capital, technological innovations and management skills? What does it mean to the global economy if there is indeed a fundamental shift in what happens in space? Can we truly look forward to what might be called an "orbital space industry" that involves solar electric power generation or new types of materials and drug production grown in a low-gravity environment? Can we eventually see true space tourism with extended stays in commercial space hotels with new types of space-based environmental protection programs? Are the people developing this new private space industry truly visionaries or simply a group out to make money from a seemingly "unique and sexy" new enterprise? Are they just pursuing "the exotic new high risk experience market" with a business model that resembles that of an entertainment park, or are they in a quest for fundamentally new technology that will unlock a new era of human experience and discovery? The last of these questions probably cannot be answered for another decade, but so far we find "all of the above" now at work in this crazy new enterprise. Dedicated engineers, space visionaries, travel consultants, crass businesspeople and scammers can all be found in the personal space industry today.

There are also serious questions about the current trend toward "weaponization of space." It is not clear how private space flight might relate to national defense, homeland security, and improved air and land transportation and navigation systems? The Bush Administration, over its eight years, resisted any attempt toward limiting options in the commercial space world as long as it did not impinge on military options. Then it was a different story. "W," Condi Rice over at the State Department, and the entire Neo-con community have vigorously resisted any suggestions, especially from Russia, that there might be a need for a new treaty to ban weapons in space. The claim from the Bush White House was that such a ban could never be verified. The key question for the new administration is what to do about space – particularly military uses of space. The Secure World Foundation, with its new head Ray Williamson seems to be intent on developing new "rules of the road" for space. This initiative will, among other things, seek to stop putting arms in space.

There are today also many serious questions about how the new spaceplane industry can be made safe from an environmental standpoint as well. How should we address the danger of debris in low earth orbit? What about rocket and jet flights in the stratosphere in terms of both the ozone layer problem and greenhouse gases? In short, there are some really key issues to consider and the space tourism industry and national air and space regulators only has about two years to resolve these problems before commercial flights are initiated for real – not just a few experimental flights that seem relatively easy to regulate today.

It is difficult to find clear precedents for this almost unique industry. It is certainly hard to find a previous start-up industry, in which governments around the world have invested, up front a significant amount of taxpayers' money. So far billions of dollars, Euros, etc. have gone into direct and indirect incentives. NASA's COTS program represents nearly a half billion dollars of public money and the people of New Mexico alone have voted to put in over a hundred million dollars in just one spaceport.

Seldom does government serve so completely as the 'launch-pad' for private enterprise. It is unusual to offer promotional support, set up an "encouraging" regulatory framework, and lend access to major governmental infrastructure. But this is the case not only in the US but also in Europe and elsewhere—sometimes on a rather grand scale. National space agencies, aviation regulatory bodies, national defense systems, missile and satellite defense systems, and the new private space initiatives are, for better or worse, closely linked together in the US and abroad.

Almost two years before its end, the George W. Bush administration announced a National Policy Directive involving access to space, defense of space infrastructure and fostering of private space initiatives. This Directive was simply announced from the White House in November 2006 and without much coordination with the US Congress. Critics have questioned the viability and effectiveness of the Policy Directive – particularly in terms of its "unilateral assertion" that the US is generally opposed to taking a "multi-lateral approach" to space. It is too early to tell whether a new administration in Washington, D.C. will alter or retain these space policies.

There is no doubt that privatization and private space ventures represent today's "big story" in the aerospace field. NASA's efforts to re-fly the Space Shuttle, complete the International Space Station and develop new vehicles that could in time go to the Moon and Mars were supposed to be the "outer space banner headlines." But NASA has been upstaged. The billionaires and the other private space entrepreneurs who are rolling out systems that promise the public sub-orbital flights in 2009 and 2010 have definitely replaced NASA in terms of public zeal – and interest. Of course one tragic accident could quickly quieten the enthusiasm for private space flight.

The last five years have seen the rapid formation of corporations to develop space planes, to market space tourism flights, develop new spaceports, and to create an interest in the possibility of "space travel." As we have seen in earlier chapters, it is Space Adventures that has led the way by arranging for Russian Soyuz vehicles to take billionaires to the International Space Station. Then the X-Prize incentives prompted a phalanx of innovative corporations to follow, not only in the US, but also around the world. Literally billions of dollars has now been invested in developing and upgrading spaceports, developing, manufacturing and testing space planes as well as in developing other needed infrastructure including hangars, launch pads, roads, testing facilities and elaborate training facilities.

This activity has quickly changed from science fiction fantasy to real worldwide business. Government space agencies have played a part in the development of this new industry, particularly outside the US. Nevertheless, many of these enterprises have eschewed government help and intervention. This is largely because the entrepreneurs felt they could move further and faster without "governmental interference and involvement."

The US Congress apparently largely agreed by turning over the licensing and regulatory control of the space tourism industry to the Federal Aviation Administration. In short, the FAA's Office of Commercial Space Transportation (FAA-AST) is in charge rather than NASA. As a result, FAA officials have undertaken an extensive "rule-making" process over the past three years and adopted regulatory controls for space plane passengers, crew and pilots. These regulations govern the "experimental stages" of the spaceplane and space tourism business. This process requires all passengers to sign waivers indicating that they hold the US Government entirely harmless against any type of accident and fully acknowledge that spaceplane flights are high risk undertakings. There are clear-cut safety controls and training standards for crew and passenger and ground control systems, but the FAA actions, in general, are geared toward encouraging the growth and development of this new industry. Passengers who "insist on going" are required to sign away their rights or those of their estate to make a claim against any accident.

The motivation provided by cash prize competitions and challenges have helped to stir the imagination of innovative aerospace engineers the world over. The success of the Burt Rutan and Paul Allen team in winning the Ansari X Prize in 2004 was undoubtedly the single most important event in the development in the space tourism business. And prizes continue to lead the charge forward. There is now the $50-million cash award being offered by Bigelow Aerospace, under the name of the "America's Challenge." This is to develop reliable and low cost transportation to Bigelow's planned space habitats.

The Google Moon X-prize at a measly $30-million has actually achieved even greater media and Internet exposure and has captured the imagination of the commercial space community around the world. This new prize is for the first commercial robotic mission to the Moon that sends back video pictures from the lunar surface as well as completing a number of other tasks. About a dozen teams have officially registered for the Google Lunar X-Prize to date (they were listed in Chapter 1). There would be others seeking the prize if Google were willing to share more flexibly in the release of the videos and news coverage that is anticipated from these private lunar missions.

In addition, there are also on offer various NASA "Centennial Prize" competitions. These are for far less money and for less awe-inspiring goals and have not captured the same amount of media coverage. Even so, this program, so energetically led by Dr. Kenneth Davidian, nevertheless could be critical to advances in space tethers, space elevators and extra-terrestrial activities.

So what will the private space industry accomplish in coming months, years and decades? What are the most important things to know about this evolving new industry? Certainly, we all have a number of questions about the future of private space flight. Well, here are some possible answers, distilled from earlier chapters. These in our view are the top ten things you should know about this curious new industry?

1. **How will the Private Space Industry be regulated and controlled in the US and Internationally?**

   The FAA Office of Commercial Space Transportation (FAA-AST) regulates this new industry under US legislation passed in late 2004 (Title 49 USC, Subtitle IX, Chapter 701) as well as Executive Order 12465. The FAA in December 2006 adopted detailed regulatory controls that were designed to ensure safety, develop appropriate licensing, training and inspection standards, and harmonize national safety and reliability standards with emerging international regulatory processes to provide oversight of a new global space tourism industry. It has also released regulatory controls for pilots and crews.

   These regulations also serve to hold the US Government harmless against accidents that may occur by requiring all passengers and crew to sign waivers against all liability and force everyone who seeks to fly into space to recognize that there are significant risks involved in space plane flights. The risks are thus believed to be much greater than riding on a commercial airplane and the regulations and waiver statements are crystal clear in this regard. The Private Spaceflight Federation (PSF) claim that if the Government will just let them be free to innovate they will come up with new vehicles and systems that are far, far safer, at least for sub-orbital flight, than anything NASA or the US Air Force has come up with to date.

   Currently the FAA licenses governmental and commercial spaceports only after thorough examinations that are conducted every five years. The FAA also licenses and authorizes, on a case-by-case basis, every flight of a commercial space plane or rocket launch and each instance is considered an "experiment." The licensing of an experimental launch is nothing like a certification of an airplane to fly multiple missions and we are many years away from such a process.

   The Congressional legislation extends this "experimental period" through 2012. This regulation is first and foremost aimed at protecting public safety. This case-by case experimental regulatory process is something that can work quite well today given the small number of spaceports and the limited number of lift-offs now occurring. In the future, however, things will change. When the number of spaceport facilities doubles or triples and the number of flights increases by a factor of ten or more, significant problems could arise. These problems could be of adequate staffing and competence by FAA-AST regulators to proceed on the basis of current case-by-case licensing processes. There are currently no particular regulations with regard to space habitats, such as those Robert Bigelow has proposed to operate. Here issues of both public safety and national security could arise.

   As described in earlier chapters, the FAA is currently placed in a somewhat awkward role by Congressional regulation and the White House Executive Order. The FAA is supposed to be regulating the industry and providing for public safety on one hand, and yet it also charged with encouraging the growth of this new enterprise on the other. Things often end badly in an enterprise where the roles of referee and cheerleader are combined.

   The former FAA's Associate Administrator for Commercial Space Transportation, Patricia Grace Smith, who left office in late 2007, made it clear that the FAA feels it will be a number of years before they will "license and certify a vehicle" as flight worthy, such as they currently do for a Boeing 777 or an Airbus 380. As Smith has said "The proven safety is simply not there."[2] Dr. George Neild, who has replaced Patty Smith at the helm of FAA-AST, has echoed the same message – safety first. Yet he knows he is also charged by Congress with promoting this new industry as the same time. The FAA is thus placed in this somewhat conflicted role.

So far the US currently leads the world in this new industry and other countries are more or less following the American regulatory formula. So far the formula has seemed to work, but once there is a real worldwide service the problem will be more clearly defined, particularly if an accident claiming human lives should occur.

It would be prudent, at least at some point in the future, therefore, for the FAA's safety and security oversight role to be divided from any responsibility to promote the industry's growth and development. Once the industry develops, it may be appropriate for their regulatory role to be restricted to only safety and security issues.

Space tourism, of course, is not just a US affair. As detailed in earlier chapters, there are many enterprises around the world seeking to develop space plane technology and systems as well as promoting spaceports in such locations as Australia, France, Singapore, United Arab Emirates, the United Kingdom, etc. The European Aviation Safety Authority (EASA) has been closely involved in two space plane development programs that are being supported by the European Space Agency. At the international level several other entities may ultimately play a role. The International Civil Aviation Organization (ICAO), and the International Telecommunication Union (ITU), are likely to be involved in regulating flight safety and the frequencies used for communications. Further, there is the newly formed International Association for the Advancement of Space Safety (IAASS) that is pursuing space safety at all levels. The IAASS is pursuing a process to create an MOU among space agencies to seek agreement on common space standards. The Secure World Foundation is trying to develop what they call international "Rules of Road" for space activities. In parallel to all the above others are pursuing the idea that the ICAO be formally designated as the entity to regulate international space operations and space plane regulation. This is an idea, however, that is far from popular among regulatory officials in the US. At this point the most popular idea is for national regulation to oversee such activities and then coordinate with other countries as necessary and appropriate.

Finally, if the development of private space planes for space tourism should branch into the development of executive supersonic jets and their safe operation, it is possible that regulation and safety controls for both enterprises can and will overlap, at least in several key areas. This would be not only in terms of FAA oversight, but also in terms of spaceport inspection and licensing, liability waivers, pilot and crew standards, safety certification and training requirements and perhaps involve some ambiguities related to insurance and risk management as well. Both the Private Spaceflight Federation and the FAA Office of Commercial Space Transportation, at this point anyway, would like to keep these two activities as separated as possible

2. How Could Space Plane and Space Tourism Safety Be Enhanced?

The US Congress should keep a close eye on FAA regulation of the space tourism industry and at some time in the future it might consider whether the safety regulatory and the licensing function should be separated by a solid wall from efforts related to the promotion of the industry.

It is also proposed that a White House Commission should be formed to address safety and security issues to define ways to advance the safe operation of the space tourism business. Just as there was recently a White House Commission that developed important new guidelines for aviation flight safety, a similar and parallel Commission might be formed to carry out this task related to space tourism. This Commission would be directed to complete its work within a specified time and be given a specific charge to carry out a number of tasks as outlined below.

- Devise ways to reduce personal space flight accidents and enhance flight safety by developing targeted and realistic quantitative and qualitative objectives and measures;
- Develop a charge for the FAA (and NASA as appropriate) in terms of creating standards for continuous safety improvement, and these goals should create targets based on performance against those standards;

- Develop improved and more rigorous standards for the ultimate certification and licensing of space planes (or their subsystems and/or their escape systems.) These should also cover their operations. This could perhaps start with certification standards and best practices for spaceports and training and simulation facilities;
- The Federal Aviation Rules for Private Spaceflight should by 2010 or so be rewritten with statements in the form of performance-based regulations wherever possible;
- The FAA should develop better quantitative models and analytic techniques to assess space plane performance and to monitor safety enhancement processes. These should be based on best industry practices as new operational vehicles and safety and emergency escape systems come on line. Key to this process is systematic input from industry with regard to flight performance for a many system and subsystem parameters;
- The FAA Office of Commercial Space and the Department of Justice should work together to ensure that full protections are in place, including new legislation if required, so that employees of the space tourism business, including but not limited to manufacturers and/or operators of space planes, maintenance and other ground based crew, owners and operators of spaceports and owners and operators of personal space flight training and simulator facilities, can report safety infractions or risk factors of concern regarding safety violations or security infractions to government officials without fear of retaliation or loss of employment – i.e. full "whistle-blower" protections for such employees. This could also include a safety and hazards reporting call-in line that is parallel to the NASA Safety Reporting System (NSRS).

The creation of the Private Spaceflight Federation is an important resource that should work with the White House Commission and the FAA to develop improved safety regulation, but a parallel entity representing space tourism passengers and crew is also needed. To this end, an entity representing spaceflight consumers should somehow provide input to the White House Commission's deliberations. This might be an offshoot of the Airline Passengers Association.

Although the first commercial space plane flights will very likely be operated by US firms or by a British firm (i.e. Virgin Galactic) operating from within the US, international regulations and licensing procedures may in time stem from other sources. Again, these include the ICAO, the European EASA organization and other professional groups such as the IAASS. These entities and others may, over time, add important safety controls and processes, but for the time being the US regulatory systems will lead the way. Let's hope they manage to get the right formula – which is actually a simple one. They are seeking to hold the reins tight enough to achieve maximum safety, but not so tight as to stifle this new industry's evolution. So far most other countries seem willing to follow the US model and are moving toward parallel or at least similar regulatory processes.

3. Who Are the Key Players in the Space Tourism Industry? And How to Follow What's Happening in This New Field?

To describe the current space tourism industry as chaotic, or somewhat like the Mad Hatter's Tea Party, would be to overstate the confusion and dramatic change now shaping this emerging field. But it is not far short of the reality. Changes, mergers, or bankruptcies are a week-to-week, if not day-to-day, occurrence. Our researches have come up with nearly fifty organizations currently in the hunt. An even more compelling statistic is that over twenty-five companies have already folded or been absorbed by other companies. (To see the diversity and the turmoil just look at Appendices A and B.) We can hope that "stability of design" will be found over time. These mature designs will most likely occur first in subsystems such as in propulsion, controls, life support systems, and environmental safety and escape systems. Some stability and standardization will ultimately be found in this new business, but certainly this can't be found in the space tourism business as it exists today. As advocates such as Diamandis, Greason, Tai and others contend, standardization today would inhibit and perhaps crush innovation.

The evolution of this new technology and industry seems in many ways a historical replay of the early days of barn-storming pilots, where everyone was building their own plane, developing their own emergency escape system and offering their own flight services to the public. Hundreds of garage mechanics and even bicycle manufacturers were trying to build the "better flying mousetrap." It was decades before the best of the airplane designs and the best of the aviation companies evolved. The same seems likely to be the case for the space tourism business. It may well be into the 2020s before this now "risky business" ultimately matures. In time there may be a significant role for lighter-than-air vehicles, advanced ion engines, space elevators, satellite-based tether systems or even technologies that have yet to be invented. It is for this reason that "type certification" of particular space planes remains a long way away. In this respect, high altitude, hypersonic corporate executive jets may in time partially lead the way. This is because they will fly longer duration flights and the potential market for those that might purchase these planes is far larger. Their safety standards may also be higher.

It is projected that a number of large aerospace companies such as Lockheed Martin, Boeing and EADS-Airbus will not only develop but offer to the market such supersonic commercial jets with noise abatement systems and as early as the early 2010s. More advanced systems such as the larger and faster A2 is perhaps decades away.

One element of seeming stability for the new space tourism industry is the newly-formed, but increasingly predominant, Private Spaceflight Federation (PSF). Its membership represents a "who's who" of the current industry – at least from the US – and it has been perhaps the main source of comments to the FAA with regard to the rule making covering the operation of space planes and the licensing of their use. Current members of the Federation are:

- Burt Rutan, President of Scaled Composites (Builder of SpaceShipOne and now the SpaceShipTwo fleet).
- John Carmack, game developer billionaire and Founder and President of Armadillo Aerospace
- Elon Musk, CEO and CTO of Space X and founder of PayPal
- Robert Bigelow, President and CEO of Bigelow Aerospace and billionaire from the hotel industry.
- Alex Tai, COO of Virgin Galactic
- Stuart Witt, General Manager of Mojave Spaceport
- Jeff Greason, CEO of XCOR Aerospace
- Gary Hudson, Founder and Chairman of AirLaunch LLC
- David Gump, President of t/Space
- George French, President and CEO of Rocketplane Ltd. (RpK)
- Eric Anderson, President and CEO of Space Adventures
- Peter Diamandis, Chairman and CEO of the X-Prize Foundation and one of the founders of the International Space University
- Rick Homans, Chairman of the New Mexico's Spaceport America and New Mexico Secretary of Economic Development (This is to be the spaceport for Virgin Galactic if the tax referendum on New Mexico State support passes.)
- Bill Khourie, Executive Director of the Oklahoma Spaceport
- Art Dula, CEO of Excalibur-Almaz

This group will undoubtedly continue to play a key role in the advancement of a space tourism business in the US as well as in the development and amendment of rules and regulations by the FAA governing safety and licensing practices. In terms of total wealth per individual member it may well be one of the most exclusive clubs in the world. The Federation will clearly help the industry find stability, share best practices and assist new entrants into the field to meet minimal standards for simulation, vehicle testing, and operational procedures. Its members will also provide the capital financing this new industry so desperately needs.

Exactly who will be the first to successfully launch a space tourism business is hard to say. Appendix A provides a more complete list of current initiatives, but the following chart, compiled by The Economist magazine (Figure 11.3) provides one view of the current leaders. But even this Economist chart is not totally complete in its list of current front-runners and the loss of the NASA development contract by Rocketplane undoubtedly has slowed its progress.[3]

There are at least a half dozen other projects that contend they will be flying in the late 2009 or 2010 time period. Only time will separate the real winners from the losers. Of course, there is more to this industry than the space plane operators. Those who provide insurance for this new industry and their success – or failure–will also be of paramount importance.

Product Launches
Leading Private Manned Spaceships

Name	Status	Finance	Seats	Launch Method	Propellant
SpaceShipOne	Completed	Paul Allen	3	Air	Hybrid
SpaceShipTwo	In development	Virgin Galactic	8	Air	Hybrid
Rocketplane XP	In development	Private / state	4	Ground	Liquid
Xerus	In development	Private investors	2	Ground	Liquid
Explorer	Design	Ansari family	6	Air	*
New Shepard	In development	Jeff Bezos	*	Ground (VTOL)†	Liquid

Source: *The Economist*   * Not known   †Vertical Take-off and Landing

One can only hope that, in time, there will be a countervailing group to the Private Spaceflight Federation to represent the views of space tourism passengers — much as there is an Airline Passenger Association today. Also professional publications will probably evolve that follow the details and key events of this new industry. Today, however, one must largely follow the development of private space flight through broader brush media such as Space.com, Space News, Aviation Week and Space Technology, by reading the many web sites of the space plane and space tourism companies, or by reviewing the official web sites of the FAA, EASA, the ICAO and IAASS or academic sites like www.spacesafety.org.

4. Is the Space Tourism Business Commercially Viable – or a Media Hyped Anomaly that Will Soon Fade from the Scene?

This is the number one question that financial analysts, insurance companies and commercial banks are asking, with a wide divergence of answers.

A few years before the dot.com bubble burst nearly a decade ago, a cynical newspaper business reporter opened his remarks at a conference in Denver by saying: "The broadband Internet commercial e-tail business has rapidly grown from a zero-million dollar activity to a zero-billion dollar enterprise." The "E-tail" business has now grown into a truly viable enterprise and is today a significant force in the global economy. But 15 years ago, it was 99+% hype and less than 1% sales. Many feel that space tourism may suffer from the same start up oversell. Serious studies have certainly been undertaken to try to understand the market for space tourism and estimate the volume of demand and the scale of this enterprise as an on-going business.

The European Space Agency has commissioned a major study and the Futron Corporation in the US undertook an in-depth study in 2002 that was substantially updated in 2006. Details were in Chapter 6. The only really hard market data, however, comes from the people who have deposited advanced booking payments with companies like Virgin Galactic to reserve a place on future flights. Altogether tens of millions of dollars in advanced booking fees have now been deposited with Virgin Galactic, Space Adventures, and others to reserve seats on future flights. This sounds impressive, but the amount shrinks to insignificance when one starts to tally up the huge investment that has now been made in spaceports and space planes. Indeed, perhaps two billion dollars have been or will likely be spent on planes, test facilities, training centers, spaceports, and space plane launcher vehicles before commercial and still "experimental" flights take off in late 2009 or 2010.

This is a capital-intensive business and a small group of billionaires have risked substantial investments in the hope that this is a venture that will "fly." (A groan can be suppressed here if you like.) A great deal of investment in space planes, spaceport infrastructure, regulatory oversight, and highly trained personnel will all be required up front, well before serious revenues start flowing.

This industry is thus a highly vulnerable enterprise that requires a lot of money be spent on insurance and risk management. A personal spaceflight launch insurance business could go south quickly if one of the commercial flights should end catastrophically at an early stage in the "take off" of the industry. It is probably true that developing a successful insurance and risk management system for the space tourism business is more important than, say, developing a better rocket engine or certainly more important than creating a highly functional spaceport.

There are lots of imponderables at this stage of the new and unproven industry's development. Just a few of these questions are:

- **What is the return or "repeat" market?**

  The sub-orbital flights involve just a couple of hours or so from launch to landing, with only about 4 minutes of weightlessness and about 10 minutes of "black sky" sightseeing. Jeff Greason's jaunt, via his XCOR-developed Lynx vehicle, climbs to a "**sub**-sub-orbital" altitude of only 37 miles (or nearly 60 kilometer). This will be a flight of about a half hour.

- **How many people will consider doing this sort of high cost flight again?**

  The answer is "no one knows." As one analyst of the space tourism ventures has suggested, this is a business that will need to develop the "and then what?" or the "what thrill next?" market. The Space Adventures people are indeed working on a number of "accessories" that they will offer to their clientele.

- **What will be the prices and the pricing curve over time?**

  The price of a sub-orbital flight is currently pegged at around $200K to $250K, (or about $100K for XCOR's lower altitude ride), but these prices will inevitably decline over time. It is thought that people with over a million dollars in assets and making $200K or so a year might be the market most likely to buy a seat on a space plane ride. Such people, who are in the upper 1% of the world's population in terms of wealth, however, are usually shrewd and intelligent buyers. In short, they got rich by investing wisely. If they are not among the first thousand to fly, it is just possible– like consumers who are waiting to buy their High Definition Plasma TVs – that they may wait until the prices fall quite a bit further. If the prices go down to say $50K there might be a new surge level of demand, and if the price falls to $20K or $25K there might be yet another surge of potential consumers. No one yet knows what the price-versus-demand sensitivity ratios will be for space tourism flights. The demand curve for a "flight to nowhere" is hard for anyone to predict. Personally we think that the low end of the curve, in terms of paying for a Space Tourism experience, may be the most successful – at least in the early years. By the "low end" we are thinking of I-Max movies, training to be an astronaut at a space center, a parabolic flight to achieve weightlessness or a flight on a very high altitude jet like the Foxbat.

- **How Many Space Plane Operators Can the Market Support?**

  Sir Richard Branson seems to be making a preemptive coup by buying a fleet of space planes from the SpaceShip Corporation that he cofounded with Burt Rutan. But Rutan has said he plans to sell as many as 40 of these vehicles to space tourism operators and, as described in previous chapters, there are many dozens of others planning to develop space planes as well. Various projected numbers for these vehicles–between 40 and 100 space planes–that might fly, frankly, do not add up in terms of a viable business plan today. It does not seem that the market can sustain so many space planes operating, at least in the t to 2012 time period. Prices can remain high only if space rides remain a very limited commodity. The high cost of buying and operating a spaceplane would appear to require high prices for some time to come. There are presumably over a million jet-setters out there who shell out big bucks for first class air fares. But the ultimate marketing answer remains elusive – only time will tell. Our best guess is that of the 40 or so spaceplane developers who say they will provide space tourism flights in coming years – only four or five will survive.

- **Will There Be Synergy Between the Corporate Supersonic Jet and Space Planes to Support Space Tourism?**

  Major aerospace companies are designing and planning to deploy "quiet" 10 to 12 passenger supersonic corporate jets that will share a number of characteristics with space planes. It is still too early to tell if the technology, the business plans, the manufacturing facilities (and scale of production) will have enough in common to help sustain both industries and also provide for their more effective regulation and control. This, however, could be a critical factor – especially in terms of sustaining a viable insurance market.

The challenges of the Supersonic Jet market for executives are dramatically different than those of the space plane entrepreneurs. Currently there are major safety, environmental and regulatory obstacles. Challenges to be overcome include the prohibition against flights in the US due to sonic booms, green house gas contrails in the stratosphere, huge developmental and operating costs, and doubts about the size of the market. The plot line and the actors are certainly quite different as well. The space plane "heroes" are the small entrepreneurs and the daring and visionary space billionaires. The script for super sonic jets calls for the giant and anonymously "gray" aerospace companies like Lockheed, Boeing, and EADS-Airbus to develop these hypersonic, high-flying new craft. Hollywood and the media thus find the space plane story a lot more compelling. Beyond the smaller supersonic business jets are the long-term and more grandiose plans like the European A2 that would fly 300 passengers, with hydrogen fueled jet engines, to speeds of up to Mach 6. The synergy, or lack thereof, between space tourism and hypersonic transport is just one of the market enigmas that potential investors are left to ponder.

5. Does a Large Number of Space Plane Flights Endanger the Ozone Layer and Present Other Environmental Dangers?

Although there are many diverse approaches now underway, virtually all efforts involve flying some form of rocket into the upper atmosphere. As is known from the experience of the Concorde SST aircraft, sustained flights high into the stratosphere can be destructive to the ozone layer and spewed out a lot of greenhouse gases at an altitude where they do a lot of harm. The ozone layer shield in the stratosphere actually protects humans and flora and fauna from intense interstellar radiation. For those who believe that preservation of humanity is actually desirable should recognize that aviation, rocket launches, and high altitude craft have been and continue to be of environmental concern. Orbital debris, ultraviolet radiation, and "killer electrons" blitzes enabled by distortions in the geomagnetic flux are also of concern.

Past experience – with jet aviation, the Concorde, etc.–suggests that the space tourism business should also be subject to some form of environmental impact process and regulatory control and that new technologies such as hydrogen propulsion systems, ion engines, space elevators and tethers, that might be less destructive to the ozone layer and could reduce greenhouse gas emissions, should be developed as soon as practicable. It also might mean that space tourism fees, and perhaps even more likely aviation tickets, could in the future be structured to include an environmental fee that would fund science or implement new environmental systems to replenish the environmental balance at the top of the stratosphere. (Certainly, the environmental issues raised by the potential depletion of, or reduction in, the ozone layer as posed by the space plane and the supersonic corporate jets are in some ways parallel – but they are largely different. The space planes fly through the ozone layer, while hypersonic transports fly along the layer for sustained periods. In both cases, however, special "sustainability" fees might be imposed on these industries. Here aviation should be expected to shoulder far more of responsibility– to address methods to help restore the ozone layer and reduce greenhouse gas emissions.)

When one recalls that the atmosphere that saves us from extinction is proportionally thinner than the skin of an apple and that the ozone layer is much, much thinner than the "shine" on the peel, the extent of our vulnerability becomes quite clear.

6. How Does The Entrepreneurial Approach To Developing New Space Systems Differ from that of NASA?

The short answer is that there are fundamentally differences. The approaches of large and "bureaucratic" governmental agencies versus agile and flexible entrepreneurial organizations are summarized in Figure 11.5 below.

7. Will the Possibility of Increased Weaponization of Space Adversely Impact The Development of Space Tourism?

In recent years there has been more and more defense-related planning that involves the possible weaponization of outer space. Current international treaties and conventions, in theory, largely prohibit the deployment of weapons in Earth Orbit or on the Moon. Nevertheless, there is a difficult distinction between defensive systems and offensive weapons. The 2007 test by China of an anti-satellite missile system that destroyed a low earth orbit satellite and exploded thousands of new debris into Earth orbit has intensified concerns in this area. The US followed in early 2008 by demonstrating its capability to destroy a "de-orbiting spy satellite." Fortunately most of the debris in the US event de-orbited in a very short time span.

As more and more satellites and space tourism facilities are deployed these questions will become even more germane. An effort to further codify a ban on weapons in space, as well as being an aid in the reduction of orbital debris, would also be of great value to those operating space tourism businesses and would also serve the interests of the peaceful uses of outer space and commercial space activities.

The National Policy Directive adopted by the Bush Administration in the Fall of 2006 is seen by some as serving to protect US assets in space and discouraging attacks on US satellites and private space initiatives. Others, however, fear that its unilateral aspects only raise the stakes and increase the likelihood of conflicts in outer space. No one knows what the next Administration in the US White House will bring and if the so-called Bush Space Doctrine will change with a new Administration. New initiatives, such as those of the Secure World Foundation to develop and help implement new "rules of road" in space, could prove very helpful if they should find international support to a set of standards and best practices that countries agree to adhere to in coming months and years.

8. How Is the Space Tourism Business Expected to Evolve?

Many have speculated that the space tourism business will need to offer a suite of new experiences to keep the supply of customers coming. Indeed one only needs to examine the web site of Space Adventures to see that a range of "space tourism" services with a broad scale of prices is already on offer. These range from a $4,000 ride on the so-called "vomit comet" in a high parabola to experience a few seconds of weightlessness, all the way up to making a $20-million trip (recently upped to $35-million) to the International Space Station for an eight-day stay. Eric Anderson, President of Space Adventures, has indicated that his company will continue to push the envelope and ultimately offer his clients the chance to engage in space walks for a mere $15-million add-on. And most recently there has been talk of a $100-million customer willing to pay for a trip on a Soyuz vehicle around the Moon.

The trick to attracting space tourism customers and to keep them in the "supply chain" is to offer new challenges, together with the incentive of fares that decline over time. Thus the "next new space thrill" may be critical to this becoming a viable market. Others suggest that if this technology is converted to actual destination travel for high-end executives then this might represent the critical path to viability.

Indeed, there are visionaries who are thinking even further ahead than Space Adventures. For example, Robert Bigelow, the hotel suite king and owner of Bigelow Aerospace of Las Vegas, Nevada, has committed his firm to deploying a space hotel. This is not a pipe dream. His prototype technology has already been demonstrated, with two test "Genesis" space habitats in orbit and sending back live video.

Nor is Bigelow alone in his belief that viable space hotels can be developed sooner rather than later. Inter Orbital Systems (IOS) of Mojave, California is developing a one-and-half stage to orbit vehicle called the Neptune (See Figure 11.7) that would convert the rocket's fuel tanks to an in-orbit habitat complete with windows to view the Earth.

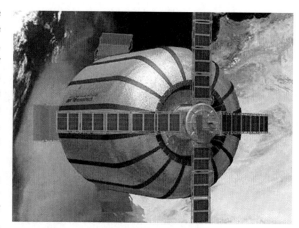

Figure 11.5: Bigelow Aerospace's Genesis 1 Spacecraft Pictured in Orbit
(Courtesy of Bigelow Aerospace)

9. What are the Possible Longer-Term Outcomes from Private Enterprise in Space?

Currently the space tourism business is focused almost exclusively on providing a unique experience to space tourists within the context of a short flight that ends at the initial departure point. John Spencer in his book *Space Tourism: Do You Want To Go?*[4] candidly observed: "…we are in the experience business. Not the space business and not the launch business…" But Spencer clearly has a very short term and indeed narrow-minded perspective of what space tourism portends. Clearly Spencer would not be a strong advocate of having an "environmental" surcharge added to space tourism flights to help sustain the planet. No one wanting to succeed in a highly profitable business wants to contend with a "downer" like paying extra to save Earth for posterity.

In the longer term, to support true space commercialization and even space colonization, planning for improved space transportation systems must ultimately follow. This is essential if private space development is truly to become a meaningful enterprise. Research and development to support entirely new systems for the safe, reliable and cost-effective creation of Solar Power Satellites (SPS), colonies on the Moon or Mars, or other longer- term commercial space activities, could and should be conceived now. And then they should be implemented as soon as possible thereafter. Hydrogen engines that produce water rather than greenhouse gases should be another priority. We need to begin critical supporting efforts, and do so now, if 21st century space programs are truly to have meaning that goes beyond a trip to an amusement park. Key infrastructure such as space elevators, advanced ion engines, mass-drivers or mag-lev devices on the surface of the Moon, satellite tether lifting devices, etc. are the inevitable follow-on to simple parabolic flights into space. Serious planning for these subsequent activities should not be "left to later" which is always the easy way out.

But in reality, is this an area in which the private sector can be expected to take the lead? Will this flirtation with space tourism and citizen astronauts be followed by another Kennedy-esque pronouncement demanding that America commits its resources to leading the world into further reaches of Outer Space? The jury is still out as to whether this is a first date, casual sex or a long term and meaningful marriage where governments and private enterprises work together to make something truly meaningful happen.

If this is seen through the eyes of a visionary such as a Diamandis or a Maryniak, the commercial space industry could be the moral and economic equivalent of the computer revolution of the 1980s. Innovation and initiative from the private sector could create new economic growth, fuel new clean energy systems, stimulate new clean aviation transport and much more. This is where creative public and private leadership is both greatly needed and unfortunately most unlikely to occur.

10. Citizen Astronauts: Reality or Science Fiction?

The race is on to make citizen astronauts a reality. More than a dozen countries are involved in various aspects of this enterprise and nearly fifty companies feel they may be able to provide a viable answer – both in terms of technology and business models. Some envision future spaceports as not only a place from which to ride into outer space but also as a venue to shop, be entertained and visit a space museum or IMAX theater. In short, they envision "excitement village." The thrilling aspect of the space tourism enterprise could be that this is a business where rocket scientists are teaming with "imagineers." Peter Diamandis, the X-Prize founder, indeed wants to host annual rocket races in New Mexico – a sort of 3D mach plus NASCAR race that grips the public imagination and spurs space commerce to new heights.

The background to all the previous chapters suggests there are many questions still to be resolved and puzzles to be solved in a host of areas. These include technology development, flight operations, safety regulation, risk management, business plans, marketing, technology transfer and intellectual property, and environmental issues coupled with new technology development and possibly "sustainability surcharges." Other key issues include weaponization of space, future uses of public and private uses of space vehicles, and developing a host of new commercial space applications and industries. Finally it raises questions about the relation between national and international governmental agencies and private enterprise in this amazing new commercial space business environment.

Some of the new space entrepreneurs (a number of whom have billions of dollars at their personal disposal) have planned test flights in the next two years and many trials are ongoing right now. The results of these will decide whether the target dates of 2009 or 2010 will be met. When the FAA reported on their Commercial Space Conference they emphasized that the leaders of this industry were saying: "No flights before their time." This is to say that the FAA and the space entrepreneurs saw themselves in accord in saying that there is no artificial timetable for commercial flights to begin. Testing and proof of concept must be complete before commercial passengers start streaming to the nearest spaceport.

Seldom in human history are the stars aligned so that so much could go wrong or so much could go right – depending on whether the right or wrong steps are taken. If private enterprise moves ahead with more sophisticated and mature ways to move people and facilities into space, very good things could happen. In time we could find ways to generate low cost and environmentally friendly electric power in space and relay it where it is needed. We could find new ways to manufacture in orbit the drugs and materials that we need to meet many societal needs. We could develop ways to build colonies on the Moon and even begin in centuries ahead to terra-form Mars so that people could live there. We could find ways to prevent comets or meteorites creating massive waves of destruction to our planet. We might even devise new tools to cope with problems related to global warming, meteorological and geological disasters and genetic mutation. Commercial space could help to make education and health care easier to distribute and reach far more students at significantly lower costs.

This could be the high road to a better human civilization and a "survivable world" for our species. The same pathway might be taken to another future that is much less desirable. Here we might find accelerated global warming, destruction of the Ozone layer, weaponization of space, and a runaway global population that is harmed rather than helped by human's newfound access to space.

The choice is ours to make. Temperate choices and smart planning can help us arrive at a future that will sustain new generations yet to come and allow our potential to be realized. Unlike Mr. Spencer in his book *Space Tourism: Do You Want to Go?*, we believe the future of commercial space travel is about much more than a cheap thrill (well actually a rather expensive thrill). The future of Space Tourism could actually turn out to be about the future of *Homo Sapiens* and the survival of the human race. Let us hope we can make the right choices and exploit this clever new technology to create a better world and sustain the human race.

References:
1. "Finding Doomsday Asteroids, *New York Times* – April 3, 2007, p. A22.
2. Speech and Questions and Answers by Patricia Grace Smith, Center for Strategic and International Studies, Washington, D.C. March 25, 2007
3. Mary Evans, "Rocket Renaissance: The era of private space flight is about to dawn" The Economist, May 11, 2006 – http://www.economist.com/science/displayStory.cfm?story_id=6911220
4. John Spencer, *Space Tourism: Do You Want To Go?* – Apogee Books, 2004, p. 38

— Appendix A —

INVENTORY OF PRIVATE SPACE COMPANIES AROUND THE WORLD

http://rocketdungeon.blogspot.com; http://home.comcast.net/~rstaff/blog_files/space_projects.htm;
http://www.spacefuture.com/vehicles/designs.shtml; http://www.hobbyspace.com/Links/RLV/RLVTable.html;
and numerous other sources as listed below.

Company	Rocket-Launch Vehicle	Intended Markets	Capabilities	Launch Site
Advent Launch Services	Advent 1 stage. (VTHL from ocean)	Sub-orbital. 300 Kg to 100 km	Full scale liquid engine tests.	Ocean launch & landing.
Aera Space Tours/Sprague Corp.	Altairis. (VTHL)	Sub-orbital. Space tourism	2 stage, RP-1/LOX propulsion. 7 passengers to 100 km suborbital flights in 2007.	US Air Force Cape Canaveral Launch Facility – 5 year agreement.
Alliant ATK	Pathfinder ALV X-1 (VTVL) http://www.astroexpo.com/news/newsdetail.asp?ID=27882&ListType=TopNews&StartDate=10/9/20006&EndDate=10/13/2006	Orbital. Launch to LEO of scientific packages. Upgradable to manned flight in time. ORS mission	Upgraded Alliant sounding rocket.	Mid Atlantic Spaceport, Wallops Island
American Astronautics	Now Renamed Sprague Corp. See Aera Space Tours. http://www.lunar.org/docs/LUNARclips/v11/v11n1/xprize.shtml	Renamed Formed to seek X-Prize. Crew to LEO. (See Aera)	See Aera above.	Cape Canaveral
Andrews	Gryphon Aerospaceplane	Sub-orbital space tourism	6360 kg to 100 km and return. LOX/RP-1. Less than $1-M/flight (in design)	N.A.
ARCA Space	Romanian project – Now Defunct http://www.lunar.org/docs/LUNARclips/v11/v11n1/xprize.shtml	Formed to seek X-Prize. Crew to LEO. Competing for Google Prize	Not available	N.A.
Armadillo Aerospace	Black Armadillo	Sub-orbital spaceflight.	1 stage. LOX/ethanol engine. (Limited capital investment). Vertical Takeoff and land. (Like Delta Clipper design.)	White Sands, New Mexico
Benson Space Company	See SpaceDev and Dreamchaser. Benson Space Company will market Dreamchaser vehicle. http://www.spacedev.com/newsite/templates/subpage_article.php?pid=583	Marketing company for Dreamchaser.	Not Applicable	Not Applicable

Appendix A (continued)

Company	Rocket-Launch Vehicle	Intended Markets	Capabilities	Launch Site
Blue Origin (Backed by Jeff Bezos)	New Shepard (VTVL) http://www.blueorigin.com/index.html	Sub-orbital. Space Tourism to 100 km.	Reusable Launch Vehicle. Hydrogen Peroxide and Kerosene fuel. Abort system.	Culberson County, Texas. HQs in Seattle, Washington
Bristol Space Planes Ltd.	Ascender (Subscale flight models). Space Bus (Concept only) 50 persons or 110 tons . Space Cab (Concept only) 8 persons or 2 + 750 Kg	Sub-orbital. 3 people or 400 Kg on space tourism flight.	Jet. 2 turbofans to 8 Km. RL-10 liquid rocket engine to 100 Km .	United Kingdom and US
C & Space (of Rep. of Korea) and AirBoss Aerospace Inc. (AAI)	Proteus space plane(VTHL) http://www.hobbyspace.com/nucleus/index.php?itemid=207	Sub-orbital. Space Tourism. 3 crew members	LOX/Methane engines. ITAR approval pending.	To be decided.
Da Vinci Program	Da Vinci (Balloon launch and vertical landing) http://www.davinciproject.com/	Sub-orbital. Space Tourism. 3 crew members	Balloon to 40,000 ft. Twin LOX/Kerosene engines to 120,000 ft. parachute landing.	Can be launched from any balloon launch site.
DTI Associates	Terrier-Orion (Terrier is surplus Navy missile motor and Orion is surplus Army missile motor)	Sub-orbital. Cargo to LEO (290 kg to 190 Kilometers)	Motors and vehicle FAA-AST licensed.	Woomera, Australia
Energia Rocket & Space Corporation	Clipper (VTOL) http://www.astroexpo.com/news/news/newsdetail.asp?ID=256888&ListType=TopNews&StartDate=5/15/2006&EndDate=5/19/2006	Orbital.	In conjunction with Soyuz and Agara Launch Vehicles.	Baikinor and Russian Northern Cosmodrome
HARC Space	Balloon Launch Reusable Vehicle	Sub-orbital. Sounding and Targeting Vehicle to sub-orbital	Balloon and liquid fuel rocket engines	Can be launched by balloon at many sites.
IL Aerospace (Israel)	Balloon launch and then Negev vehicle to Sub-orbital space http://web1-xprize.primary.net/teams/ilat.php	Sub-orbital. 10 km by balloon and then Negev rocket launch to 120 km.	Balloon and Negev solid fuel rocket with parachute to water landing	Can be launched by balloon at many sites. Israel base.
Inter Orbital Systems (Mojave, California)	Sea Star (13 Kg to LEO) Neptune (4500 Kg to LEO) http://www.interorbital.com/	Sea Star. Microsat Launch Vehicle.	Stage and a Half. Liquid bipropellant rocket. FAA-AST licensed	Off shore. Pacific Ocean. Los Angeles and Tonga
Japanese Aerospace Exploration Agency (JAXA)	HII Transfer Vehicle (Unmanned but in time might be upgraded to manned and pressurized vehicle.)	Unmanned Cargo resupply to the ISS. Launched on the HII vehicle.	Conceptual studies	To be decided.

Appendix A (continued)

Company	Rocket-Launch Vehicle	Intended Markets	Capabilities	Launch Site
Japanese Rocket Society	Kankoh Maru (Latest version of earlier Phoenix design)	Orbital. 50 passengers to 200 km LEO.	Single Stage to Orbit. Vertical Takeoff	No hardware designs.
JP Aerospace (Rancho Cordova, Cal 95742)	Access to Orbit-Ascender Balloon System (Not this is different system than that of the Bristol Spaceplanes. http://www.jpaerospace.com	Sub-orbital. High Altitude experiments or rocket launch.	Very High Altitude Balloon. Can be used as Launch Platform to LEO using ion engines	California sites
Kelly Space & Technology Inc. (San Bernadino, California)	Space plane http://www.kellyspace.com/	Crew and Satellite and Cargo launch. sub-orbital	Tow launch of reusable space plane	San Bernadino Airport
Lockheed Martin-EADS	Autonomous Transfer Vehicle (ATV) (Unmanned but could be upgraded to manned and pressurized vehicle.)	Unmanned Cargo re-supply to the ISS. Launched Ariane 5 but can also be launched on Atlas 5.	Could be upgraded to become a manned vehicle. Not yet funded.	Atlas 5 site at Cape Canaveral.
Lorrey Aerospace (Grantham, NH 03753)	X 106 Hyper Dart Delta http//www.lorrey.biz/	Orbital. Pilot + passenger and 220 Kgs. To LEO or Bigelow space station. Orbital data haven.	(Conversion of F 106 Delta Dart to include ramjet to create a spaceplane.)	To be decided.
Masten Space	XA 1.0 (VTVL) XA 1.5 (VTVL) XA 2.0 (VTVL) http://www.masten-space.com/products.html	Sub-orbital. XA 1.0 100 Kg to 100 km, XA 1.5 200 kg to 500 km, 2000 kg (5 people) to 500 km.	Liquid reusable internalized engines.	To be decided.
Planet Space (See also Canadian Arrow)	Silver Dart Spaceplane and Lifting Body (VTHL) http://www.thestar.com/NASAApp/cs/ContentServer?pagename=thestar/Layout/Article_Type1&c=Article&cid=1155678611503&call_pageid=968332188492 http://www.planetspace.org/lo/index.htm Canadian Arrow http://www.canadianarrow.com	Orbital. Crew of 8 Sub-orbital. Crew of 3	First stage liquid propellant + OX. Second stage 4 JATO rockets-Abort	Nova Scotia, Da Vinci Spaceport

Appendix A (continued)

Company	Rocket-Launch Vehicle	Intended Markets	Capabilities	Launch Site
Rocketplane-Kistler	K-1 (5700 Kg to LEO, 900-1400 Kg to GTO) Falcon Rocketplane XP Pathfinder	Orbital. Payloads to LEO, MEO, GTO, ISS Cargo re-supply & return missions. Sub-orbital. Cargo & Microsats. Sub-orbital. 4 seat fighter-sized vehicle. Up to 4 or 410 kg to 100 km. Or microgravity experiments	Various propulsion systems for K-1, Falcon and Rocket plane XP Pathfinder.	Woomera, Australia and Nevada Test Site for K-1
Scaled Composites-Spaceship Corporation -Virgin Galactic (Mojave, California)	SpaceShipTwo-SS (HTHL)	Sub-orbital. Space Tourism 7 people to 100 km	Neoprene & NO2 as oxidizer.	Mojave Airport, South West Regional Spaceport (SRS)
Space Adventures with Myasishchev Design Bureau & Federal Russian Space Agency	Explorer Space Plane (C-21) and MX-55 High Altitude launcher plane (HTHL)	Sub-orbital. Space Tourism	Liquid fuel motors. Horizontal Takeoff and Horizontal Landing (lifting body with parachute landing)	To operate from a number of international spaceports including Dubai, Singapore, US et al
SpaceDev (California)	Dreamchaser (VLHL)	Sub-orbital. Space Tourism (1 stage) 6 passengers Orbital (2 stage manned access to ISS)	Single Hybrid Engine. (Neoprene and NO2) for sub-orbit. Launch of spaceplane on the side of 3 large hybrid boosters to reach LEO orbit & ISS.	To be decided.
SpaceHab (Contract agreement with NASA to support COTS)	Apex 1 Apex 2 Apex 3 http://www.astroexpo.com/news/newsdetail.asp?ID=27197&ListType=TopNews&StartDate=8/21/2006&EndDate=8/25/2006	Orbital. Launch to LEO orbit. (300 kg (Apex 1) to 6000 kg (Apex 3) Apex 1&2 unmanned. Apex 3 can be manned.	Open architecture to support different missions and NASA's COTS Program.	To be decided.
Space Transport Corp. (Forks, Washington)	Rubicon 1 &2 and N-SOLV now to be replaced by Spartan vehicle.	Sub-orbital. 2 passengers to 80-100km. Spartan can launch 5 kg to LEO.	Design and status of project, and financing not clear	To be decided.
Space Exploration Technologies (Space X)	Falcon 9, Dragon Space plane http://www.spacex.com/	Orbital. Commercial Orbital Transport Service to ISS	Cluster of 9 Merlin engines on Falcon 9.	Kwajalein Atoll launch complex
Starchaser Industries (UK and Rocket City New Mexico)	Thunderstar-Starchaser 5	Sub-orbital. Space Tourism. Launch to 60 km.	Bi-liquid. LOX & kerosene rockets. Parachute recovery.	To be decided.

Appendix A (continued)

Company	Rocket-Launch Vehicle	Intended Markets	Capabilities	Launch Site
Sub-Orbital Corp. and Myasishchev Design Bureau	M-55X and Cosmopolis XXI	Sub-orbital. Two stage to 100km. Pilot and 2 passengers. Space Tourism.	1st Stage M-55X Geophysika. 2nd stage. C-21 a rocket-powered lifting body with parachute landing.	Flexible launch and takeoff sites.
Transformation Space Corp. t/Space (Allied with Scaled Composites)	CXV (Crew Transfer Vehicle)	Orbital. Crew of 4 to LEO or ISS & ISS re-supply missions	Launches at high altitude from a large cargo carrier aircraft.	To be decided.
TGV Rocket	Michelle B Rocket (Modular Incremental Compact High Energy Low cost Launch Experiment)	Sub-orbital. Small crew or scientific instruments.	Single Stage to orbit. Modular.	White Sands, New Mexico
Triton Systems	Stellar-J (HTHL)	Orbital. 440 Kg of cargo to LEO	Launches via a cargo jet and LOX-Kerosene	To be decided.
UP Aerospace	SI-1 Carrier Rocket-Space Loft XL.	Launch of small scientific packages of 50 Kg in 220 km LEO Orbit.	Liquid fueled rocket. (Licensed by FAA-AST)	Spaceport America, New Mexico
Vela Technologies	Spacecruiser	Sub-orbital. Space plane. Up to 8 people	Jet plus Propane/NO_2	To be decided.
Wickman Spacecraft & Propulsion	WSPC Small launch Vehicle, and SHARP Space plane (VTHL) http://www.space-rockets.com/sharp.html	Orbital. Cargo and in time Crew to LEO. Eventually spaceplane to carry passengers.	Phase Stabilized Ammonium Nitrate solid fuel rocket. 900 kg to LEO.	To be decided.
XCOR Aerospace	Sphinx (Sub-orbital space) (HTHL) Xerus (Sub-orbital space) (HTHL)	Sub-orbital. Space Tourism and nanosatellite launch. Xerus can also launch 10kg microsatellite to LEO orbit.	Isopropyl alcohol/LOX Sphinx is FAA/AST licensed.	White Sands, New Mexico

— Appendix B —

Other Space Plane Initiatives Thought Now To Be Defunct

Sources include: http://rocketdungeon.blogspot.com; http://home.comcast.net/~rstaff/blog_files/space_projects.htm; http://www.spacefuture.com/vehicles/designs.shtml; http://www.hobbyspace.com/Links/RLV/RLVTable.html

A.	Acceleration Engineering This company proposed to develop the Lucky Seven single-stage manned sub-orbital vehicle. (Defunct former X-Prize candidate).
B.	Aero Astro LLC. Designed the PA-X low-cost expendable launch vehicle. (Defunct former X-Prize candidate).
C.	Applied Astronautics. – This company proposed to develop a low cost reusable sub-orbital launch vehicle – the Hyperion.
D.	Beal Aerospace Technologies. Inc.–Initiated development of the BA-2 low-cost heavy-lift expendable launch vehicle.
E.	Blue Ridge Nebula. This was a small family enterprise initiated to try to win the X-Prize. It is now defunct
F.	Canyon Space Team - Proposed to develop the XPV sub-orbital reusable space vehicle for space tourism flights.
G.	Discraft Corp. This company proposed to develop a 'blastwave-pulsejet-powered' space vehicle. (Defunct former X-Prize candidate).
H.	E-Prime Aerospace Corp. This initiative planned to turn a MX Peacekeeper ICBM into the so- called Eagle launch vehicle.
I.	Fundamental Technology Systems. They proposed to develop the Aurora reusable manned space plane. (Defunct former X-Prize candidate).
J.	Kitty Hawk Technologies. (formerly Cerulean Freight Forwarding Co). designed the Kitten reusable manned sub-orbital space plane concept. (Defunct former X-Prize candidate).
K.	Lone Star Space Access Corp. (formerly Dynamica Research designed the Cosmos Mariner sub-orbital space plane concept. (Defunct former X-Prize candidate).
L.	Micro-Space Inc.. This company proposed to develop the Crusader X manned sub-orbital space-sled. (Defunct former X-Prize candidate).
M.	Microcosm Inc. This company actively sought to develop the Scorpius family of low-cost expendable launch vehicles.
N.	Panaero Inc.. (formerly Third Millennium Aerospace Inc..) This group proposed to turn a Sabre 40 jet into the SabreRocket, a rocket-powered manned reusable sub-orbital space plane. (Defunct former X-Prize candidate).
O.	Pioneer Rocketplane This company designed the Pathfinder refuelable manned space plane. (Defunct former X-Prize candidate).
P.	Platforms International Corp. This company sought to design the SpaceRay space plane for sub-orbital space tourism flights.
Q.	Pogo. This company sought to design a reusable first stage for Spacelift using Aircraft Jet Engines and Vertical Take-off and Landing technology– designed by Glenn Olson.
R.	Rocket Development Company. This group designed the Intrepid expendable launch vehicle for Universal Space Lines.
S.	Rotary Rocket This company proposed to develop the Roton single-stage-to-orbit manned reusable launch vehicle that lands under a rotor. This project eventually collapsed in 2000. (Defunct former X-Prize candidate).
T.	Sky Ramp Technology. This now defunct company had sought to develop a sub-orbital space plane.
U.	Space America Inc. This company sought to develop the Enterprise 4 low-cost heavy-lift expendable launch vehicle with a capsule landing.

V.	Space Access. This unit planned to develop the SA-2 ramjet powered space plane for the sub-orbital space tourism business.
W.	The Spacefleet Project. This company sought to develop not only the SF-01 sub-orbital tourist vehicle, but also the SF-01B orbital vehicle to reach low earth orbit.
X.	Starcraft Booster Inc. (This company was chaired by moon walker Buzz Aldrin) It proposed to develop a reusable winged booster staged rocket system.
Y.	Thrugate Aerospace. This company sought to develop a sub-orbital space plane.
Z.	Truax Engineering Inc. This group sought to design the Excalibur family of amphibious reusable launch vehicles.
AA.	Universal Space Lines LLC. This group planned to develop the Spaceclipper sub-orbital space plane.

A 50-YEAR HISTORY OF SPACE TOURISM: CHRONOLOGY OF EVENTS IN THE EVOLUTION OF COMMERCIAL SPACE FLIGHT

(with other key dates in space development)

Date	Event
1957 October	Launch of Sputnik by the USSR – the world's first artificial satellite.
1958 November	President Eisenhower created NASA – a reaction to the Soviet space program and the desire to launch a US satellite.
1961 April May	Soviet cosmonaut Yuri Gagarin is the first man to orbit the earth. First Mercury manned space flight of the US space program.
1962 February	John Glenn is the first US astronaut to orbit the earth.
1965 June	Gemini-4 astronauts undertake the first EVA (i.e. spacewalk)
1967 June Summer	Three astronauts killed in Apollo launch-pad accident. Barron Hilton, President of Hilton Hotels published a paper about space tourism – "Hotels in Space."
1968 April	Premiere of "2001, A Space Odyssey" – Academy Award winning movie by Stanley Kubrick and Arthur C. Clarke which included scenes of a manned space station and lifestyle in Outer Space.
1969 July	First Apollo moon landing; Neil Armstrong is the first man on the moon.
1972 December	Last manned space flight to the moon; Gene Cernan is the "last man on the moon."
1977 May	Skylab launched by NASA.
1981 April	First flight of the Space Shuttle.
1982 Spring	British Aerospace start development of HOTOL – Horizontal Take-off and Landing spacecraft (canceled due to lack of funding in 1986).
1984 November	First of a series of papers on SSTO and HTOL vehicles for space tourism published by David Ashford of Bristol Space planes (UK).
1985 August	Study on Space Tourism by Society Expeditions of Seattle presented to NASA and the L-5 Society Space Development Conference. Design of "Phoenix" SSTO-VTOL passenger vehicle published by Gary Hudson of Pacific American Launch Systems. NASA started design work on HL-20 astronaut escape vehicle (canceled in early 90's after costs of $2-billion)
1986 January October	Loss of Space Shuttle Challenger with seven astronauts. "Potential Economic Implications of the Development of Space Tourism" presented at IAF Congress, including estimate of market.
1987 October	Design of passenger-carrying upper stage for German "Stenger" SSTO-HTOL vehicle presented at IAF Congress by Dietrich Koelle.

Appendix C (continued)

Date	Event
1988 September	Space Shuttle flights resumed 2.5 years after the loss of Challenger.
1989 October	Design for an orbital hotel presented in "Feasibility of Space Tourism" session at IAF Congress by Shimizu Corporation, a major construction company.
1990	NASA funded development of the Rockwell X-30 National Space Plane (NASP) – a Single Stage to Orbit (SSTO) spacecraft (work was terminated in 1993).
1991 November	McDonnell Douglas started work on the Delta Clipper DCX reusable launch vehicle (canceled in 1996 after 8 test flights) International Space Conference of Pacific-basin Societies (ISCOPS) in Kyoto received papers on Phoenix ("History of the Phoenix VTOL SSTO and Recent Developments in Single-Stage Launch Systems") and Space tourism ("Benefits of commercial passenger space travel for society.")
1992 October	IAF Congress considered "The Prospects for Space Tourism: Investigation on the Economic and Technological Feasibility of Commercial Passenger Transportation into Low Earth Orbit" by Sven Abitzsch and Fabian Eilingsfeld. Starchaser Industries founded in the UK to develop Britain's role in the space industry.
1993 April July	Japanese Rocket Society started a study program on the feasibility of space tourism and established its Transportation Research Committee to design a passenger launch vehicle. SpaceHab made first flight on board Shuttle Endeavor.
1994 March May May October	American Society of Civil Engineers (ASCE) "SPACE 94" conference in Albuquerque considered the feasibility of commercial "space business parks." International Symposium on Space Technology and Science (ISTS) in Yokohama, received paper from the JRS study program on Kankoh-Maru, the JRS passenger launch vehicle to carry 50 passengers to and from LEO. The Commercial Space Transportation Study Final Report included the first study by major US aerospace companies of the potential market for space tourism – coming to the conclusion that space tourism wasn't feasible(!) The design of Kankoh-Maru presented at the annual IAF Congress; and a 1/20 scale model of Kankoh-Maru displayed at Farnborough International Air Show.
1995 Spring September September October	NASA JSC started development of X-38 Lifting Body followed by a contract with Scaled Composites Inc. (project canceled in April 2002 due to budget pressures). The Space Transportation Association (STA) in Washington DC started a study of space tourism with cooperation from NASA. Market research on the demand for space tourism in Canada, the USA, and Germany showed a huge potential market worldwide. US Office of Commercial Space Transportation (OCST) formally moved into the Federal Aviation Administration.
1996 January March/April May May June July July	NASA started development of X-43A and X-43C Space Planes, a $250-million program (terminated in late 2004 after the third test flight). Cover story of Ad Astra, the magazine of the US National Space Society, is "Space Tourism" – for the first time. "X" Prize project launched at a Gala Dinner in St. Louis. Speakers (including NASA Administrator Dan Goldin) linked the "X" Prize to space tourism. At the 20th ISTS in Japan, NASA's Barbara Stone presented "Space Tourism: The Making of a New Industry" which concluded: "Studies and surveys worldwide suggest that space tourism has the potential to be the next major space business." American Society of Civil Engineers "SPACE 96" conference considered the legal issues that need to be resolved before private commercial facilities can be constructed in orbit. NASA announced award of $900-million 3-year contract to Lockheed-Martin to build and fly the X-33 un-piloted, reusable rocket test-vehicle to speeds of Mach 15 (canceled in March 2001) STA-NASA space tourism study steering group concluded that the obstacles facing establishment of a space tourism industry could be overcome "within 15 years."

Date	Event
1996	
August	Contract awarded to Orbital Sciences Corp. for development of X-34 test RLV – Resusable Launch Vehicle (canceled in March 2001 after 3 test flights).
September	California Spaceport became the first commercial US spaceport to be licensed by the FAA-AST.
November	Aerospace America article ("Japan plans day trips to space") on Kankoh-Maru is the first mainstream aerospace journal coverage on the subject.
1997	
February	IEEE Aerospace Conference at Snowmass, Colorado, featured three papers on space tourism
March	STA-NASA workshop in Washington DC considered a range of issues relating to establishing space tourism business.
March	First International Symposium on Space Tourism held in Bremen, Germany, organized by Space Tours GMBH.
April	Aviation Week published article "Studies claim space tourism feasible" based on papers presented at the IEEE Aerospace Conference (February, above) – the first time they had covered the subject.
May	The FAA/AST issued the Spaceport operated by the Florida Space Authority a license to operate. (This was renewed for five years in 2002)
July	Cheap Access To Space (CATS) conference held in Washington DC jointly sponsored by NASA and the Space Frontier Foundation.
October	International Astronautical Federation (IAF) President Karl Doetsch referred to space tourism as "one of the only businesses which will enable the launch industry to grow significantly"
November	Announcement of "Space Tourism Society" based in LA, chaired by Buzz Aldrin.
December	FAA/AST issued a launch site operator's license to the Mid-Atlantic Regional Spaceport (MARS)
1998	
January	AIAA workshop in Banff, Canada on international cooperation in space included Space Tourism as one of five themes and recommended that "in light of its great potential, public space travel should be viewed as the next large, new area of commercial space activity."
February	Business Week, Fortune and Popular Science all published articles on the US venture companies Kelly Space Technology, Kistler Aerospace, Pioneer Rocketplane and Rotary Rocket that were developing reusable launch vehicles.
March	Press conference on Capitol Hill announced the release of "General Public Space Travel and Tourism," the final report of a joint study by STA and NASA started in September 1995. NASA admitted that space tourism is both feasible and economically desirable.
March	NASA Administrator Goldin in a speech at NASA 40th anniversary gala dinner said: "...in a few decades there will be a thriving tourist industry on the Moon."
April	First modules of the International Space Station (ISS) launched.
April	Formation of Bigelow Aerospace Inc. announced.
April	Space 98, biennial space conference of ASCE in Albuquerque had sessions on space tourism, space commercialization, space access and space ports, among many others.
May	The X Prize Foundation announced target of $10-million and launched the "X Prize" for the first commercial company to demonstrate a reusable passenger space vehicle.
May	FAA started a study for extending air traffic management upwards to include low Earth orbit.
September	Space Policy Journal published article, "Space Tourism: a response to continuing decay in US civil space financial support."
October	Papers on space tourism featured at European Space Agency (ESA) workshop on Space Exploration and Resources Exploitation.
1999	
July	Contract awarded by NASA to Boeing Co. to develop the X-37 Approach and Landing Test Vehicle (ALTV) with a 4-year program valued at $173-million.
2000	
January	Norman Augustine, ex-CEO of Lockheed-Martin Corporation, predicted in 'Aviation Week' that space tourism would become the main space activity.
January	Formation of Mircorp announced to commercialize the MIR space station.
March	Illustrations of the Japanese Kankoh-Maru and Bristol Spaceplane's' "Spacebus" appeared on the NASA web site.

Date	Event
2000	
June	ISTS Symposium in Japan included space tourism sessions on Universal Spacelines, the X-Prize, airline operations, insurance, and certification of Kankoh-Maru for passenger carrying.
June	MirCorp announced the first fare-paying guest to visit MIR is Dennis Tito, founder of Wilshire Associates.
June	2nd annual conference in Washington DC of the Space Travel and Tourism Division of the Space Transportation Association.
July-August	International Space University (ISU) Summer Session included Design Project on Space Tourism (for the first time).
October	First meeting on space tourism in France at the French space agency, CNES. Presentations by CNES, ESA, Astrium and ISU.
2001	
February	Conference in Washington DC sponsored by the US Federal Aviation Administration entitled "The Prospects for Passenger Space Travel"
March	NASA canceled X-33 and X-34 RLV technology demonstrator programs after spending over $1-billion and encountering numerous technical problems.
April	Dennis Tito became the first paying space tourist, launching from Baikonur aboard a Russian Soyuz bound for the International Space Station. Tito returned safely after 128 orbits in 8 days.
June	Space Exploration Corp. (SpaceX) announced a program to develop the Falcon series of 2-stage launch vehicles.
June	NASA funds research into whether US citizens would like to take a trip into space. The survey confirmed the potentially huge market for space tourism and led to the conclusion that only space tourism offered a large enough market to enable reusable launch vehicles to reduce the cost of getting to orbit.
June	First US Congressional hearing on space tourism at the House subcommittee on Space and Aeronautics.
July	NASA released "General Public Space Travel and Tourism" – the very positive report on feasibility of space tourism originally produced by NASA in 1998.
July	Rocketplane Inc. formed to develop plans for space vehicles.
August	Mircorp announced agreement with the Russian government and RSC Energia to design, develop, launch and operate the world's first private space station, Mini Station 1.
October	2nd IAF Congress in Toulouse, France included several symposia dedicated to space tourism and other possible new space markets.
October	XCOR Aerospace successfully completed first phase of its flight test program for the EZ-Rocket - the world's first privately built rocket-powered airplane.
November	Space Adventures, Ltd., commissioned market survey on space tourism. The market analysis stated that, at the price of $100,000, more than 10,000 people per year would purchase flights.
November	X-Prize competitor Starchaser Industries successfully launched its Nova single-seat sub-orbital rocket for first time. The un-piloted test from Morecambe Sands, England, reached 1688.8 meters (5541 feet).
November	The Presidential Commission on the Future of the US Aerospace Industry held its first hearing. Astronaut Buzz Aldrin, one of the Commissioners, argued the case for the importance of developing "high volume human space transportation."
2003	
February	Loss of Space Shuttle Columbia with seven astronauts.
October	China launched Shenzhu 5 and astronaut Yang Li Wei to become the third nation to successfully launch a manned spaceflight.
2004	
January	President George W. Bush announced new space vision "to explore the moon, Mars and beyond."
September	SpaceShipOne team of Paul Allen and Bert Rutan won the Ansari X-prize for commercial space flight.
September	Sir Richard Branson of Virgin Atlantic and Bert Rutan, who developed SpaceShipOne, announced the formation of Virgin Galactic to provide passenger flights into space "in 2.5 to 3 years."
	Virgin Galactic reached a 20-year lease agreement for use of a New Mexico spaceport
December	US Congress passed the Commercial Space Launch Amendments Act of 2004 (CSLAA) which makes the Department of Transportation and the Federal Aviation Administration (FAA), responsible for regulating human space flight.

Date	Event
2005	
January	Jeff Bezos announced plans to create a spaceport facility to support his Blue Origin launch operations of the New Shepard vehicle in Van Horn, Texas.
September	NASA announced broad design features for Project Constellation.
October	Interorbital Systems announced first "Promotional Fare" spaceline ticket to Tim Reed of Gladstone, Missouri.
December	NASA issued RFP to industry for commercial services to ferry supplies and astronauts to the ISS - the COTS scheme (Commercial Orbital Transportation Services).
December	Virgin Galactic reached a 20-year lease agreement for use of a New Mexico spaceport
2006	
January	The FAA published "Human Space Flight Requirements for Crew and Space Flight Participants; Proposed Rule" in the Federal Register. It contained recommended requirements for crew qualifications, training and notification, as well as training and informed consent requirements for space flight participants.
February	Rocketplane Inc. purchased Kistler Aerospace, forming RpK.
February	Space Adventures announced Explorer and Xerus Space planes and Spaceports in Singapore and the United Arab Emirates
July	Bigelow Aerospace successfully launched Genesis-1, prototype for an orbital hotel with a target date of 2015.
August	The FAA and US Air Force Space Command issued new common federal launch safety standards designed to create consistent, integrated space launch rules.
August	SpaceX and Rocketplane Inc. awarded a contract by NASA to provide the K-1 vehicle to ferry personnel and supplies to the International Space Station under the COTS scheme.
August	Planetspace/Canadian Arrow announced permanent spaceport facility in Cape Breton, Nova Scotia
September	Anousheh Ansari became the 4th paying passenger (and the first woman) to fly into orbit with Soyuz, spending two days on the International Space Station.
September	SpaceDev Corporation announced plans for Dreamchaser, a passenger carrying spacecraft based on NASA's HL-20 design.
September	Virgin Galactic announced that flights with 6 passengers to reach a sub-orbital altitude of 140 km will cost $190,000; reports of "100's of reservations" being made, including celebrity names.
September	Space Shuttle flight program was resumed 3.5 years after the loss of Columbia, to continue construction of the International Space Station. A further 15 Shuttle flights are planned.
2007	
April	Charles Simonyi from Los Angeles became the 5th fare-paying passenger on Soyuz flight to ISS.
June	Astrium announced European space vehicle at Paris Air Show
June	NASA Space Shuttle Atlantis carried out successful mission to the International Space Station
July	Bigelow Aerospace launched Genesis 2 inflatable module from Russia.
July	Space Adventures offered a spacewalk on future Soyuz mission for $15-million in addition to $30-million flight to ISS.
July	Aerospace giant Northrop-Grumman increased its stake in Scaled Composites from 40% to 100%.
July	Space Adventures announced cirumlunar mission on Soyuz spacecraft with two places at "$100-million per couch"
July	First private space industry fatalities as three are killed and three injured in engine testing explosion at Scaled Composites facility in Mojave Desert.
July	Armadillo Aerospace announced successful tethered tests at its New Mexico base.
September	Google announced $30 million Lunar X Prize for first private moon rover vehicle to send images and data to Earth.
October	NASA dropped Rocketplane Kistler from the COTS program for failing to meet financial goals.
October	China launched its first lunar probe to begin a 3-phase program to land an astronaut on the Moon by 2020
October	John Carmack's Armadillo Aerospace narrowly failed in attempt to win Northrop Grumman's $350,000 Lunar Lander challenge.

Date	Event
2007	
November	Planetspace announced its proposal to compete for the NASA COTS award.
December	Sir Richard Branson completed 2-day training program at NASTAR Center in preparation for his inaugural flight on SpaceShipTwo.
December	Richard Garriott, 36, named as sixth passenger to fly to ISS on board Russian Soyuz, in October 2008.
2008	
January	Virgin Galactic unveiled new models of SpaceShipTwo and WhiteKnightTwo in New York. NASA announced $4.7 million dollar contract to Zero Gravity Corporation to provide weightless flights to NASA-operated experiments and personnel.
February	The X PRIZE Foundation and Google announced the first ten teams to register for the $30-million Google Lunar X Prize.
	Orbital Sciences Corp. won $171-million NASA award to build and demonstrate a launch system capable of delivering cargo to the international space station under its $500 million COTS program (replacing RpK).
March	XCOR announced the Lynx, a two-seater spaceship smaller than a private jet to take people up for a 25-minute space flight.
July	Space Adventures and RSA (Russia) announced first Soyuz trip to the ISS with two passengers, scheduled for 2011.
	Scaled Composites rolled out the WhiteKnightTwo mothership at Mojave CA spaceport to prepare for test flight program.
August	SpaceX suffered third failure of its Falcon launch vehicle on the first flight with a payload of 3 small satellites.
	SpaceDev signed contract with Scaled Composites for development of the rocket motors for SpaceShipTwo.
	Patricia Grace Smith, former Assoc. Administrator for Commercial Space Transportation at FAA joins the Board of SpaceDev.
September	Chinese astronaut makes the country's first spacewalk during the third manned spaceflight.
	Space X achieves first successful flight of Falcon 1 which became the first privately-developed rocket to reach earth orbit.
October	Richard Garriott was the sixth fare-paying passenger to fly to the ISS on board a Soyuz spacecraft.

— Appendix D —

GLOSSARY OF TERMS

A2: Design concept for a hypersonic transport, as developed for the European Space Agency.

AADC: Alaska Aerospace Development Corporation

AATE: Argentine Association for Space Technology

ACS: Attitude Control System

ACTS: Advanced Crew Transportation System

AIAA: The American Institute of Aeronautics and Astronautics

ALPA: Air Line Pilots Association International

ALTV: Approach and Landing Test Vehicle

Ares 1: The vehicle being developed to lift crew to low earth orbit

Atlas: The U.S. developed launch vehicle. The heavy lift configuration is currently the Atlas V.

Ares 5: The heavy lift vehicle that will lift payload to low earth orbit to mate with the crew for missions to the Moon under Project Constellation.

Ariane: The launch vehicles operated by Arianespace. Currently the Ariane V operates from the Kourou launch range in French Guyana.

ASE: Association of Space Explorers – USA

ASRI: Australian Space Research Institute

ATV: Automatic Transfer Vehicle of the European Space Agency. Designed to carry cargo to and from the International Space Station. The first ATV is named the Jules Verne.

Blue Origin: Blue Origin, LLC – the company developing the New Shepard launch vehicle.

BM: Booster Module, as proposed by Inter Orbital Systems, Inc. for their Neptune launcher.

CASC: The Chinese Aerospace Corporation

CFR: Code of Federal Regulation

CM: Crew Member

CM: Crew Module as proposed by Inter Orbital Systems, Inc. for their Neptune launcher.

CNSA: Chinese National Space Agency

CNES: National Center for Space Studies, the French Space Agency.

CO_2: Carbon dioxide gas.

Constellation: Also known as Project Constellation. The U.S. project to send Astronauts to the Moon and eventually to Mars. The Ares and Orion crafts developed under this project can also be used to reach the International Space Station.

COPUOS: The Committee on the Peaceful Uses of Outer Space of the United Nations

COTS: The Commercial Orbital Transportation System

CSA: The Canadian Space Agency

CSIS: Center for Strategic and International Studies

CSLAA: Commercial Space Launch Amendments Act of 2004

CSTLR: Commercial Space Transportation Licensing Regulations

DARPA: The United States Defense Advanced Research Projects Agency

Delta: The U.S. developed launch launch system. Currently the majority of launches are provided via the Delta II configuraton.

DOD: United States Department of Defense

DOT: Department of Transportation of the United States

EADS: European Aerospace Defense Systems, this Company was created through the merger of British Aerospace and Matra-Marconi of France.

EASA: The European Aviation Safety Authority

ECLSS: Environmental Control and Life Support System

EES: Emergency Escape System (EES)

EPA: Environmental Protection Agency

EPS:	Electric Power System (EPS)
ESA:	European Space Agency
EU:	European Union
EURECA:	The European Retrievable Carrier
FAA:	Federal Aviation Administration of the United States
FLPP:	Future Launch Preparatory Program of ESA
Genesis:	The name of the inflatable "test" space habitats deployed by the Bigelow Aerospace Corporation
GIG:	Global Information Grid
GLIN:	Global Legal Information Network
H2 and H2A:	The Japanese launch systems developed by JAXA.
HTHL:	Horizontal Take-Off and Horizontal-Landing
HTVL:	Horizontal Take-Off and Vertical Landing (Note: Vertical landing from this take-off mode usually implies a capsule landing by parachute either on land or in the ocean.)
HTV:	The Japanese transfer vehicle designed to carry cargo to and from the International Space Station.
IAASS:	International Association for the Advancement of Space Safety
IADC:	The Inter-Agency Space Debris Coordination (of the UN)
IASE:	International Association of Space Entrepreneurs
ICAO:	International Civil Aviation Organization
ILAT:	IL Aerospace Technologies (Israel)
IOS:	Inter Orbital Systems, Inc.
IR:	Infra Red
ISS:	The International Space Station
ISLAP:	Institute for Space Law and Policy
ISRO:	The Indian Space Research Organization
ITAR:	International Trade in Arms Regulations of the U.S. Government. This controls the export of sensitive and strategic U.S. technology.
ITU:	The International Telecommunication Union
JAXA:	Japanese Aerospace eXploration Agency (JAXA)
JRS:	Japanese Rocket Society
LiOH:	Lithium Hydroxide
LNG:	Liquid Natural Gas
LOX:	Liquid Oxygen
LSS:	Life Support System
MARS:	Mid-Atlantic Regional Spaceport
MDB:	Myasishchev Design Bureau (Russia)
MIRV:	Multiple Intercontinental Re-entry Vehicle
MPL:	Maximum Probable Loss
MUOS:	Mobile User Operational Satellite System
NASA:	National Aeronautical and Space Administration of the United States
NASCAR:	National Association for Auto Stock Car Racing
NASTAR:	The National AeroSpace Training and Research Center
NATO:	North Atlantic Treaty Organization
NEO:	Near Earth Object
NO_x:	Nitrous Oxides of various types that are emitted by some solid fuel rockets
NPRM:	Notice of Proposed Rule Making – (a term used by U.S. Independent Regulatory Commissions such as by the Federal Aviation Administration.)
OAS:	Organization of African States
Orion:	The craft designed to carry crew to orbit and provide a splashdown capsule for crew as part of Project Constellation.
OSC:	Orbital Sciences Corporation
OSHA:	Occupational Safety and Health Administration
OMB:	Office of Management and Budget

OSM: Orbital Station Module, as proposed by Inter Orbital Systems for their Neptune launcher.
OSMA: Office of Safety and Mission Assurance (NASA)
OTV: Orbital Test Vehicle
Planehook: Planehook Aviation Services, LLC
PRA: Paperwork Reduction Act of 1995.
Prodea: Prodea, LLC The company owned by the Ansari family.
PSF: Personal Spaceflight Federation
Regulations, The: Commercial Space Transportation Licensing Regulations
RpK: Rocketplane-Kistler Inc.
RLV: Reusable Launch Vehicle
SAAHTO: Stage and A Half To Orbit Launch Vehicle
SDI: Strategic Defense Initiatives (also known as "Star Wars")
SHARE: Satellites for Health and Rural Education (Intelsat project)
SSTO: Single Stage to Orbit Launch Vehicle
Space X: Space Explorations Technologies Corporation (SpaceX)
SPF: Space Flight Participant
Taurus: The launch vehicle that Orbital Science is modifying to serve as the key vehicle to support the COTS Program.
TGV: TGV Rockets, Inc.
TSAT: Transformational Satellite System
TSTO: Two-stage-to-orbit vehicle
UAV: Unmanned Autonomous Vehicle
USAF: The U.S. Air Force
VTVL: Vertical Take-Off and Vertical Landing
VTHL: Vertical Take-Off and Horizontal Landing
t/Space: Transitional Space Corporation
UN: The United Nations. Sometimes referred to as the United Nations Organization (UNO).
USAF: United States Air Force
XCOR: XCOR Aerospace Inc.
X Prize: The non-profit organization that creates prizes to stimulate the development of new and innovative technology. (This began with the Ansari X Prize of $10 million awarded for the first successful spaceplane demonstration in 2004.)

Author's Note:
Please see Appendices A and B that also provide the names of many commercial spaceflight entities

— Author Biographies —

Joseph N. Pelton

Dr. Joseph N. Pelton holds a Ph.D. from Georgetown University and is Director of the Space and Advanced Communications Research Institute (SACRI) at George Washington University. He is also the founder of the Arthur C. Clarke Foundation and founding President of the Society of Satellite Professionals International. He also served as Dean and Chairman of the Board of Trustees of the International Space University, Director of the Interdisciplinary Telecommunications Program at the University of Colorado at Boulder, and Director of Strategic Policy at Intelsat, he is also the former Executive Editor of the *Journal of International Space Communications*.

His awards include: Outstanding Educator award of the International Communications Association, the H. Rex Lee Award for public service of the Public Service Satellite Consortium, and the ISCe Award for Outstanding Educational Achievement. He received the 2001 Arthur C. Clarke Lifetime Achievement Award and was also elected Chairman of the Academic Committee of the International Association for the Advancement of Space Safety (IAASS) of which he is a Fellow. He has made a number of media appearances on US television and radio, BBC radio CBC of Canada. Dr. Pelton is the author of hundreds of articles, and twenty-five books in the fields of telecommunications, and space policy and systems, including the Pulitzer Prize nominated book *Global Talk*. This book also won the Eugene Emme Literature award of the American Astronautics Society.

Peter Marshall

Peter Marshall began as a journalist with BBC radio and television news, later becoming a News Editor. He then joined Visnews of London (now Reuters-TV), a global TV news agency where he became General Manager. He was a pioneer in the use of satellites for TV news coverage and distribution and in 1986, he moved to Washington DC to create the Broadcast Services Division of the inter-governmental body INTELSAT. With the emergence of deregulation and competition, he returned to the private sector as President of Keystone Communications. This organization, now part of Globecast (a subsidiary of France Telecom) is the largest provider of global satellite broadcast services.

Peter Marshall is a past Chairman of the Royal Television Society in the UK, past President of the Society of Satellite Professionals (SSPI) in the USA and he was elected to the SSPI's "Hall of Fame" in 2006 in recognition of his pioneering work for the industry. He serves as a member of the Board of the Arthur C. Clarke Foundation. Peter Marshall is now a writer and consultant, based in his native UK, and he has collaborated with Dr. Pelton on five books and major research projects, including *Communications Satellites – Global Change Agents* (published by Lawrence Erlbaum in 2006) and *Space Exploration and Astronaut Safety* (published by AIAA in 2007).